The Bump

Battle of Britain 40th Anniversary Reunion
Royal Air Force Biggin Hill
21st September 1980

by Nicholas Wright

Nicholas Wright
September 1980

Designed by Berney Baughen
Photographic copying by Christopher Mercer

First published by RAF Biggin Hill 1980

Photographs:
The photographs in this book are from a variety of sources and have been selected to give representative illustration of the subject matter. In cases where the quality of reproduction is poor, the photograph has been included as the best available illustration of an important subject. Where known, photograph sources have been accredited.

Cover illustration:
An oil painting by David Shepherd commissioned by the station officers in 1953. This untitled painting, depicting two Hurricanes attacking Me109s, was David Shepherd's first venture into aviation painting. The original hangs today in the Officers' Mess at Biggin Hill.
(Reproduced by kind permission of David Shepherd)

Typeset in Garamond series by ABC photocomposition, Downe, Kent
Printed by CW Printing Limited, Borough Green, Kent

ISBN 0 9507218 0 8

The Bump

Contents

Foreword

Wing Commander J R Myers LLB RAF
Officer Commanding
Royal Air Force Biggin Hill

Every year, we commemorate Battle of Britain Sunday at St George's Chapel of Remembrance, Royal Air Force Biggin Hill. As we finalised plans for last year's Service, we realized that 1980 would be the 40th Anniversary of the Battle and that a special effort should be made to mark the occasion; for this most famous fighter station is undoubtedly best remembered in the minds of the British public for its vital contribution to the defence of our country during the summer of 1940.

That is how the idea of a reunion of personnel who served at the station during the period July to October 1940 was born; the 1980 station would thus play host to the 1940 personnel. We were determined to track down not only the aircrew and ground staff of the seven squadrons based here during the Battle, but also station staff, who included WAAFs, Army personnel, and many civilians.

For the Battle of Britain was not won in the air alone. The aircrew will for ever enjoy the just distinction of having been 'The Few', but the victory was equally shared with many others. On the ground were the mechanics, the armourers, the drivers, the cooks and so on, the devoted WAAFs whose discipline and spirit were an inspiration, the Women's Land Army, the Ack-Ack and the Balloon Commands, personnel of the NAAFI, the Salvation Army, the WVS, the Police, the Fire and Ambulance Services. Also, there were the war workers and civilian population who cheerfully endured the fire of the enemy, hundreds being killed or wounded on airfields, in factories, and in their homes, as the pilots fought above their heads. Those few pilots fought against great odds, but they had the heart and the hands of the whole nation behind them.

We do not pretend for one moment that the Battle of Britain was won at Biggin Hill. Many other stations were involved and achieved as much, but the name of Biggin Hill will for ever convey the glory of those momentous days, with a victory as decisive and significant as the Armada, Trafalgar or Waterloo. One of my most famous predecessors, Group Captain 'Sailor' Malan, summed up the vivid fighting history of his station:

"It is the fighter station with the best spirit and the best performance due, I think, to good teamwork on the ground and in the air – and it is a happy station."

And a viewpoint of a station officer in the later days of the war still holds good today:

"*. . . discipline and teamwork made the men that made Biggin a tradition, and that same Biggin tradition has shaped and formed the men who have gone to it since those early days.*"

Although fighter aircraft no longer operate from the station, we are all very conscious of the important role it continues to play in today's Royal Air Force as the home of the Officers and Aircrew Selection Centre.

In welcoming you back to the *Bump*, I know that many of you will think back to those dark days of 40 years ago with nostalgia, and even with some regret for the spirit and unity our nation seems to have lost somewhere since. But I am sure that no one wants to go back to wartime; we are best left with our memories, and I am sure that we will never forget.

For the British people that great conflict in which the Royal Air Force repulsed the onslaught of the German Luftwaffe has left the name of the Battle of Britain proudly adorned with a lustre which time will never tarnish. "If the British Empire and its Commonwealth last for a thousand years, men will still say, '*This was their Finest Hour*'."

The Station History

Biggin Hill has been connected with military aviation for more than sixty years. Its name became illustrious during those summer months of 1940, but 1940 was just a brief period in a long career of distinguished service. The following pages outline the story of the station from its foundation in 1916 to the present day.

The Battle of Britain is highlighted in some detail as a separate section which divides the History into two distinct halves.

1916-1940

Early Days

Biggin Hill became a military airfield in the latter part of the First World War. Today, as a military installation it is some 63 years old, and for more than half of these years – from early 1918 until January 1958 with just a few gaps – it was the home of front-line fighter units.

The airfield owes its establishment to the development of wireless telephony in the Royal Flying Corps and the need for the Wireless Testing Park at Joyce Green, on the Thames near Dartford, to expand and find a location capable of accommodating its increasingly heavy workload.

In the summer of 1916, Major Orme, the CO of the Testing Park, instructed his staff to search for alternative sites, but suitable locations were scarce and a potential new home was not found until a Lieutenant Hansard, who was from Limpsfield, a small village nestling under the North Downs on the Surrey-Kent border, remembered that on the high ground above his home there were many sizeable fields on the open heathland.

Hansard and a colleague motored south from London to investigate. South of Bromley, they came to a large level field, 80 acres or so, sitting above a steep valley which would make it relatively fog and mist-free. The field had good turf and nearby was a large empty country house which would make an excellent mess. There was plenty of room for building and a reasonable road, leading to Bromley and to London, passed nearby. The site was at Leaves Green (Biggin Hill), and it was ideal.

On investigating the large house, Hansard and his colleagues were challenged by a caretaker who informed them that the field was already designated as a night emergency landing ground and that he was retained by the War Office to oversee security and preparation.

So the site already had official military connections. The War Office and the General Staff had been reluctant to appreciate the importance and potential of wireless telephony in military aviation, but by late 1916 they had been converted and they duly authorised the relocation of the cramped Testing Park from Joyce Green. The Biggin Hill site was selected and land acquisition was set in progress. Most of the land was on Cudham Lodge Farm, tenanted by John Westacott from Lord Stanhope, the owner of Chevening and the Cudham Lodge Estate. Stanhope had previously accepted the occasional use of his estate as the night landing ground, and he quite readily agreed to the disposal of further land – despite the breaking-up of John Westacott's farm.

The large house – *Koonowla House* – however, posed something of a problem. It had been bequeathed with a large endowment to the Victoria Hospital for Sick Children in Chelsea, and although the building was not in use, the endowment was valuable and the trustees were loath to give it up. They could not be persuaded, and eventually the Defence of the Realm Act had to be invoked by the War Office and the house requisitioned.

The Wireless Testing Park Moves In

With acquisition finally settled, the Wireless Testing Park spent the closing weeks of 1916 preparing their new location and getting ready for the move, levelling bumps and filling hollows, erecting the canvas hangars, making workshops and offices – many of them from aircraft packing cases. *Koonowla* was used as the Officers' Mess, whilst the NCOs and men were accommodated under canvas or in temporary wooden huts. The first aircraft was scheduled to arrive on New Year's Day 1917, but was delayed by fog at Joyce Green and came on the 2nd, a day late. It was an RE7 piloted by Lieutenant Dickie, with Air Mechanic Chadwick as passenger. On touchdown it was greeted by a fusillade of snowballs as it parked in front of the hangar. The crew then climbed down from the cockpit and kissed the frozen airfield soil in solemn ceremony.

By mid-February, the transfer was complete and the Wireless Testing Park was able to continue its important work of research and development. A priority task was the perfection of a practical and operational system of air-to-air telephony to specifications laid down by Lord Trenchard.

This was to prove a vital advantage to the RFC Squadrons engaged against the Germans above the trenches in France.

The Air Defence of London

The development of heavier-than-air military aviation during the First World War had been left largely to the Army, whilst the Navy tended to concentrate on balloons and airships. The Army had devoted most of its efforts to the operational squadrons in France and little thought had been given to the aerial defence of the United Kingdom. Meanwhile, the Germans has been using Zeppelin airships as bombing platforms and were causing increasing casualties and damage in English towns, particularly in London and the South-East. The provision of air defences had not really been seriously considered and depended in the main on inadequate and ill-organised mobile guns and searchlights. The frequency and intensity of the German raids increased as they switched from Zeppelins to Gotha long-range bombers, and the casualties mounted. It became necessary to divert fighter aircraft from the trenches and form a defensive force to combat these raids. During 1917 Parliament commissioned General Smuts to investigate urgently the whole question of organised air defence, and arising from his report came the immediate plan for an air defence force covering the London area, and a longer-term proposal for an independent air force – later to become the Royal Air Force.

At first, there were only six squadrons available for London defence but gradually the force was built up and squadrons were established at locations to the south and east of London. Results were immediate, Zeppelin raids virtually ceased and the Gotha bombers were switched from daylight raids to night attacks. The limited range of the Gothas dictated that they took the direct route to London and this brought them close to the Wireless Testing Park at Biggin Hill. As the summer of 1917 gave way to the long nights of autumn and winter, the prospect of Gotha raids increased, and the two additional fighter-defensive squadrons were authorised – 140 and 141 – the latter to be based at Biggin Hill.

141 Squadron demonstrate the wonders of air-to-ground wireless telephony at Sutton Farm in 1918.

39 and 141 Squadrons

It was early winter and the authorised formation date was not until 1 January 1918, for it would take at least until then to man and equip a new fighting unit. Meanwhile, South London could not be left undefended so 'D' Flight of 39 Squadron was despatched from Hornchurch with BE2es and BE12s to plug the gap. 39 Squadron arrived on 1 December 1917 and set up a tented camp on the north-west side of the airfield, above Salt Box Hill, well away from the Wireless Testing Park. Thus 'North Camp' became established.

39 held the fort through the freezing winter, making a number of uneventful and fruitless sorties against raiding Gothas, and on 8 February 1918 the newly-formed 141 Squadron moved in from Rochford to join forces with the detached flight of 39, and establish itself on the *Bump* as Biggin Hill's first operational squadron.

By this time a start had been made on more permanent buildings on North Camp and the Wireless Testing Park had expanded to become the Wireless Experimental Establishment – WEE for short. The pilots of 141 were an international group, Australians, Canadians, Rhodesians, New Zealanders, even a Punjabi; they were very young and many of them were being 'rested' from the rigours and pressures of flying on the Western Front. They had the boisterous high spirits of a group of young men drawn together by a common bond in unusual and trying conditions. A friendly but very keen rivalry quickly developed between these fighting men of 141 and the 'elderly boffins of WEE', and under the leadership of Major Philip Babington (later Air Marshal Sir Philip Babington), 141 was moulded into a very effective fighting unit with a high esprit de corps.

The First Kill

Nevertheless, real action was hard to find and the Squadron's early months were spent working with WEE on wireless development and with the Royal Engineers on improving the techniques of aircraft location. Their equipment was now the Bristol Fighter, but it was not until 18 May 1918 that 141, under the command of Major B E Baker (later Air Marshal Sir Brian Baker), and a unit of the newly-formed Royal Air Force, was able to put this machine through its paces in the squadron's first successful action against the Hun.

On this night the Germans had mounted their biggest raid of the war, amassing more than forty bombers, mainly Gothas, against London. 141 was 'scrambled' and was led to the raiders with reasonable accuracy by the primitive sound detector system. Lieutenants Turner and Barwise picked up a Gotha over Downe and moved their Bristol Fighter in to attack. They scored direct hits with tracer and machine gun, but the Gotha, although mortally wounded, with one engine 'out' and streaming fuel, did not crash and was lost to its pursuers as it side-slipped into cloud. It was then to be found and bagged by a Major Sowrey (father of Air Marshal Sir Frederick Sowrey) of 143 Squadron, who was patrolling in his Camel and finally downed the stricken Gotha over Harrietsham. Later investigation proved that the first attack had delivered the fatal blow, and 141 and Biggin Hill had thus achieved their first victory.

Less than half of the forty German raiders reached their target that night, and eight of them were brought down. The Germans accepted that the London defences were more than a match for their bombers and no further raids were mounted.

This meant a quiet life for the squadrons of the London Air Defence Force, and boredom through enforced inactivity began to set in. In September, in order to regenerate a sense of pride and purpose, the GOC organised a Squadron-at-Arms competition. All units in the Command were to participate in tests involving formation flying, aerobatics, wireless telephony, gunnery, and a Concours d'Elegance of machines, airfield and buildings. The winning squadron would receive a silver cup, the title 'Cock Squadron', and the right to display the fighting cock emblem on their aircraft. At first, 141 Squadron were reluctant entrants but they picked up the competitive spirit as they succeeded in the eliminating rounds, and they came out on top as 'Cock Squadron' in a very keenly fought final at Suttons Farm in Essex.

Pilots of 141 Squadron celebrate in style as 'Cock Squadron' in the Squadron-at-Arms Competition 1918.
(Photo: Imperial War Museum)

After the War

The Armistice soon followed and brought with it a period of prolonged uncertainty. Demobilisation was slow and inefficient, morale was low, and the nation was disappointed with the peace that failed to bring the promised land of milk and honey. Almost five years of the 'War to end all Wars' had created a backlash calling for disarmament – pacifism was the vogue and the future of the infant Royal Air Force looked shaky indeed.

141 Squadron departed for Southern Ireland in March 1919 and was replaced on the *Bump* by 37 Squadron, who came from Stow Maries with Sopwith Camels and were renumbered '39' on 1 July. To reduce development costs, the WEE was reorganised into a central research establishment, combining four previously separate units under one roof. All research and development, other than into engines and airframes, was to be undertaken at this new unit under the title Instrument Design Establishment (IDE).

39 Squadron stayed for two years before disbanding in 1921, leaving the IDE in sole possession until the arrival of 56 Squadron, who came with Sopwith Snipes in May 1922.

56 Squadron, The Army and The Night Flying Flight

Trenchard fought a valiant battle in defence of his protegy air force, and the

Salisbury Report of 1923 secured its independent future. It established Biggin Hill as a permanent location with prime responsibilities for the air defence of London. IDE moved out to Farnborough (later to become the Royal Aircraft Establishment), and Biggin Hill settled into its long-term peacetime role as a fighter station. The Army retained a very active interest in the site, with their anti-aircraft and aircraft detection development units moving in to the South Camp accommodation vacated on the move of the IDE. The station thus remained two camps, the Army at the southern site and the Air Force to the north.

That is not to say there was division and non co-operation, quite the reverse, and for some years the role of 56 Squadron was to be closely associated with the development of R/T and aircraft detection. Also, a small RAF unit, the Night Flying Flight, remained on South Camp, working with their cousins in khaki, their role being exclusively that of aircraft detection, and their equipment being three Vickers Vimys.

The 1920s can best be described as a period of consolidation during which the Air Force found its feet and established itself as an independent service with its own systems, customs and traditions.

56 Squadron re-equipped with Gloster Grebes in September 1924 and continued in the relatively dull world of a fighter squadron in peacetime – repetitive training and exercises, gunnery camps, army co-operation practice etc – but they also had the R/T development flying to keep them busy.

By 1927 expansion had been authorised and Biggin was to become a two-squadron station. To facilitate this an extensive building programme had to be undertaken and North Camp was chosen as the area for permanent extension; the temporary hangars and workshops, and the hutted accommodation were all to be replaced with handsome brick buildings.

'Concrete Radar'

56 Squadron re-equipped with Siskins in the autumn and then moved out to North Weald, leaving the *Bump* ready for building, and with the Night Flying Flight on South Camp as the only remaining RAF unit. The work of the Night Flying Flight and the Army Air Defence Unit was absorbing and of considerable importance – even though the progress of other research was later to turn it into something of a 'blind alley' of development. By the late twenties the unit had developed a system of huge acoustic concrete dishes and mirrors, to be placed strategically along the south and east coasts. 200 foot long concave strips were to alternate with huge circular discs. They bore an uncanny resemblance to the later radar dishes which were of course to supersede them before the concrete 'Eyes and Ears' system was ever put into practice. Even so, a few of the dishes were constructed and installed for development purposes, and some can still be seen along the Kent coast today, peering seawards like giant fossilized radars.

North Camp is Built

Development of North Camp was delayed somewhat as various pockets of additional land were purchased and a further large chunk of Cudham Lodge Farm was acquired as essential extension to the airfield. Building started in the summer of 1929 and by this time the schedule included South Camp, the Night Flying Flight having gone, leaving the station empty, bar a small maintenance staff. Double-section Belfast hangars were put up, new workshops, a stores complex, station headquarters, barrack blocks, the large three-storey Officers' Mess, the Sergeants' Mess and Airmen's Mess, married quarters and the host of small buildings that make up a whole station, were constructed in the three years ending in autumn 1932. These are the buildings which stand on North Camp today – the hangars were destroyed in the raids of 1940 and the barrack blocks and married quarters were damaged, but the majority of the station as it stands today was built in the period 1929 to 1932.

Halcyon Days

The new camp was vacated by the builders in September 1932 and the station reopened under the command of Wing Commander G B Dacre. The first 'pair' of squadrons was formed by 32 with Bulldogs, and 23 with Bulldogs and Hawker Harts. They moved in across the valley from Kenley, leaving the Surrey station to be rebuilt in the fashion of Biggin.

Wing Commander E O Grenfell took command early in 1933 and stayed for four years, whilst 23 and 32 Squadrons quickly settled in to peacetime life on the *Bump*. 23 re-equipped with the new Hawker Demon and the two squadrons were regular participants in the annual Hendon Air Pageants. In common with most other UK units, Biggin Hill also opened its gates to the public on the yearly Empire Air Days and the resident squadrons always ensured that an unusual and exciting show was laid on.

Bristol Bulldogs of 32 Squadron.
(Photo: John Topham Picture Library)

The end of 1935 brought a significant change in the European political situation and, in consequence, to the life-style of the squadrons at Biggin Hill. In October, Mussolini invaded Abyssinia and a large section of 23 Squadron was shipped out to Malta as part of a strategic reserve in case hostilities should spread. 23 did not return to Biggin, for whilst in Malta their detached flight expanded to give birth to 74 Squadron, and the remnant aircrew moved on from the *Bump* to reform in full squadron strength at Northolt in December 1936.

Preparation for War

Meanwhile, the situation on the home front was rapidly changing. As 1936 progressed, the nation became more aware that war was a real possibility, and effort, finance and resources were at last devoted to armament – especially to re-equipping and building-up the Royal Air Force. Air Defence in Great Britain was reorganised

Empire Air Day 1934.
(Photo: John Topham Picture Library)

Hawker Demon of 23 Squadron
(Photo: John Topham Picture Library)

into four functional commands: Bomber, Fighter, Coastal and Training, Biggin Hill becoming one of the foremost stations in the new Fighter Command. Radar, or Radio Direction Finding – RDF – as it was then (secretly) called, was progressing at a pace, and from the late summer of 1936 Biggin was very much involved in putting the theory and potential of this development into practical effect.

32 Squadron worked unceasingly, putting in many hundreds of flying hours as an unofficial 'fighter development unit', establishing and perfecting techniques and procedures for the conversion of early warning knowledge into effective fighter counter measures. Operations Room systems were developed using the simple 'Principle of Equal Angles', enabling controllers to plot invaders' exact positions and courses, and to direct intercepting fighters straight to them. In consequence of this early work the station was for a while a school for fighter controllers. 32's Bulldogs were replaced by Gauntlet IIs, and then in March 1937 'B' Flight was separated to form the nucleus of 79 Squadron. 79 then quickly received additional pilots and machines to train-up and operate alongside its clone parent in serious preparation for war. For a while the two squadrons continued to work with their Gauntlets, but on Empire Air Day in 1938 a single Hurricane Mk I was on static display, whilst a Spitfire – the first and only at that time – staged a spectacular overhead demonstration, moving on to repeat its circus performance at five other stations.

On the Eve of War

The Munich appeasement of September 1938 brought the station to war readiness, with one squadron always on standby. 601 (County of London) Auxiliary Squadron at Hendon was partly mobilised and sent a detachment of Demons to join 32 and 79 and become familiar with Biggin as their designated wartime location. All aircraft were painted in drab camouflage and 32 received a Miles Magister in which to train for the Hurricanes which began to arrive in October.

By late November, both 32 and 79 Squadrons were re-equipped with Hurricane Is. The station was now under the command of Wing Commander Richard Grice and was making ready for war. Trees and shrubs were planted, and the runways and aprons were toned down with paint and bitumen to disguise them from the air, trenches and shelters were dug, windows reinforced, blast walls and sandbag reinforcements were constructed, and Army Ack-Ack and ground defence units were moved in to give cover from attack. But this preparation did not go unobserved, for the daily Lufthansa (at that time a civil airline only in name) flight between Croydon and Berlin sometimes passed low over the station to record and photograph progress.

Meanwhile, the two squadrons worked-up to full operational readiness in their new Hurricanes, and perfected the system of fighter control and direction. They were joined by 3 Squadron from Kenley between May and September, who came

79 Squadron Hurricane Is running up outside the North Camp Belfast hangars on Empire Day 1939.
(Photo: Kent Messenger)

Biggin Hill from the south – August 1939. This is a reconnaissance photograph taken at low-level from a Lufthansa civil aircraft on a scheduled flight between London (Croydon) and Berlin. The old South Camp shows clearly in the foreground with *Koonowla House* in the centre. The row of terraced married quarters to the right is today called *Crossley Close*, after Squadron Leader Michael Crossley. North Camp is top left.
(Photo: Ministry of Defence)

for experience on Hurricanes, and then 601 Squadron moved permanently from Hendon, arriving with Blenheims temporarily adapted as fighters by the addition of a box of four Browning machine guns under their bomb-bays. With the declaration of war in September 1939, Biggin Hill was ready – 32, 79 and 601 Squadrons and Army Ack-Ack units being present in force.

The 'Phoney War'

War, at first, came as something of a relief – the long months of training and tense preparation were at last to be converted into action. The bubble had burst and the young fighter pilots would now be able to do their bit. But action was not to be found. The station settled into the Phoney War of the winter of 1939-1940. Patrols were routine and uneventful, and such 'scrambles' as did occur were either false alarms or abortive.

79 Squadron broke the lull by making the station's first kill on 21 November, when two of their Hurricanes, operating from the forward airfield at Hawkinge, brought down a Dornier 17 over the Channel, off Dover. This was followed on 30 November by a very successful action involving six of 601 Squadron's Blenheims, operating briefly out of Northolt and jointly with six Blenheims of 25 Squadron, in a surprise raid on the Luftwaffe seaplane base at Borkum in the Friesian Islands. Five seaplanes and some fast patrol boats were totally destroyed without any loss by the raiding Blenheims. But these actions apart, life was generally very quiet and December brought a return to routine patrols without further incident or excitement.

In January 1940, advantage was taken of the 'lull' and the three squadrons were dispersed whilst a much-needed concrete runway and deep shelters were constructed. 32 went to the *Bump's* satellite airfield at Gravesend, 79 went to Manston and 601 to Tangmere. 601 was not to return, but 32 and 79 were back again in April, resuming once more their routine patrols over the Channel.

France and Dunkirk

Routine patrols were not to last, however, for by now the war situation was rapidly changing. Germany invaded the Low Countries on 10 May, and 79 Squadron was dispatched immediately to Merville, in France. Within hours, 79 was replaced by 610 (County of Chester) Auxiliary Squadron, which came from Prestwick with Spitfire Is to become the first Spitfire unit to operate from the *Bump*.

32 Squadron saw its first action on 11 May in an unfortunate raid on the Dutch airfield at Ypenburg. This field had been reported as taken by the Germans in their advance through Holland, and 32 was ordered to make a harassing straffing attack. They completed this attack with some efficiency, only to learn on their return that they had attacked Dutchmen who had temporarily regained control of the airfield. They were depressed by this disaster, but events did not allow them to dwell on

their misfortune, for the series of Army failures on the Continent meant that they and 610 Squadron were both on busy patrols over Northern France, two or three times daily, fighting a raging air battle against huge odds in a valiant attempt to clear the skies above the desperate Allied armies on the ground.

But the odds proved overwhelming and as the armies retreated towards Dunkirk, Air Force units based in France were pulled back, some in disarray. 79 Squadron returned as evacuees after less than two weeks fighting – very severe fighting – for in this period they moved base three times and claimed twenty-five enemy aircraft, but at the loss of many of their pilots and most of their precious Hurricanes which were left bombed and burned-out on the airfield at Merville. At this time Biggin Hill doubled as a front-line fighter station and troop reception unit, and in the space of a few days received more than 800 airmen evacuated from the crumbling resistance to the German onslaught. Meanwhile, the battle raged relentlessly on and there was no respite for the three Biggin Hill squadrons – even the shattered 79 went straight back into action on its retreat from France. By late May the British armies were huddled on the Dunkirk beaches and 'Operation Dynamo' – the plan for their evacuation by sea – was put into effect. 32 and 79 were exhausted, their men and machines were spent and they were stood down to rest and regroup – 32 to Acklington, and 79 to Digby. Their immediate replacements were 213 and 242 Squadrons, both with Hurricane Is.

213, the 'Hornets', were well-experienced, having been blooded in France alongside 79; for 242, however, this was to be their first action – a real baptism of fire. 610 continued until 4 June when it was pulled out to rest at Gravesend. By this time they had eight kills and four probables but they had lost six Spitfires – half their squadron – and they could not have continued without needless sacrifice. Their replacement was 229 who came in with Hurricanes from Digby.

By the end of the first week of June 'Dynamo' was over. The British Army had been snatched from defeat by the Royal Navy and by a flotilla of little boats. The operation had been made possible by the Royal Air Force whose constant patrols in the skies above the beaches had kept the Luftwaffe at bay, thus enabling the evacuation to proceed. All the RAF units in the South East had played their part, and Biggin Hill had earned its first Battle Honour. In nine days of fighting the station had accounted for thirty-six German aircraft, with a further twenty probables, and a loss of eight pilots and fourteen of its own aircraft.

But there was no respite – no time to lick wounds or to rest on laurels – for the Luftwaffe had consolidated its position in France and in the Low Countries, and it now turned its attention to Britain, and in particular to the RAF. The Dunkirk replacement squadrons departed and 32 and 79 – Father and Son – returned to their familiar home. The tempo and intensity of Luftwaffe raids increased steadily through June and July – The Battle of Britain was about to begin.

Their Finest Hour

"If the British Empire and its Commonwealth last for a thousand years, men will still say, 'This was their Finest Hour'."

No Lull Before The Storm

With the fall of the Low Countries and then of France, Hitler had control of Continental Europe. Behind him to the east lay the sleeping giant of Russia, and ahead to the west, across that tantilisingly narrow stretch of water, lay the tempting islands of Britain. Hitler was not prepared at this stage for adventures against Russia, but he needed Britain, and with the British Army in disarray following its rout in France, and with America showing no wish to become involved, he knew that he needed to strike while the iron was hot. That time was now, for Britain stood alone.

Very quickly, Hitler amassed his armies along the French and Belgian coasts and put together an invasion fleet in preparation for a massive landing. He knew that he could succeed where no one had dared to try for close on nine-hundred years, but first he needed clear skies above his army's heads. Before his invasion could be launched he needed to wipe out the Royal Air Force, and this had to be done quickly so that Britain would be his before the onset of winter.

The Germans were now operating out of the French airfields used barely a month earlier by Fighter Command squadrons. They amassed a formidable array of fighters along the Continental coast from Brittany to Norway, and further inland were their bombers. In all, it was an immense force of some 3,500 aircraft on 400 airfields. Against them, spread on airfields between Fife in the north and Cornwall in the south and west, were Fighter Command's fifty-seven squadrons totalling a mere 600 or so aircraft. But every day more Spitfires and Hurricanes came off the production lines and more young pilots, many of them Polish, Czech, French, Dutch and Belgian, as well as English, Irish, Scots, Welsh and Commonwealth, emerged from the desperately short training machine to join our gallant squadrons. In so many cases the transition was from sixth-form school desk to parachute pack in a fighter cockpit in just a few short weeks. Every day counted, for every extra day was of infinite value in bringing down those daunting odds.

For a while the Luftwaffe's tactics were cautious – they were testing the water before taking the plunge, and they confined their activities to small but frequent 'hit-and-run' raids on the coastal towns of Kent and the South Coast, and to shipping passing close to the English shores. In this way they were able to draw Fighter Command, and by using only small elements of their total force they caused our fighters to move to the forward airfields of Hawkinge, Lympne and Manston, where they were tied to being constantly on the alert and the defensive – a tiring and nerve-racking battle of attrition.

Hitler could not have realised that even at this early stage Fighter Command was already stretched almost to its limits, and if he had launched his 'Adler Tag' – Eagle Day – against the RAF in those days of late June or early July he would almost certainly have succeeded. But he did not – his intelligence must have been inadequate for it is clear that the Germans assumed the aircraft and pilot strength of Fighter Command to be far greater than it was.

For Biggin Hill these were wearying days of repeated defensive patrols and 'scrambles' in reaction to the coastal raids. It meant flying down to Hawkinge early morning, operating from the coast all day, and returning to Biggin to sleep a fitful night – and it meant repeating this day after day after day. Scores were low but the effect on men and machines – our men and machines – was serious for we were getting battle-weary before the battle had started.

The race was on and we were desperate to gain as much ground as possible against the Hun. Every possible marginal improvement in technique and tactics was exploited to the full and the manufacturers were busily up-rating and modifying the already excellent machines. Late in June all the Hurricanes and Spitfires were fitted with constant speed (variable pitch) conversion units to give them a better rate of

27 June 1940. HM King George VI holds an Investiture on the Station. Flight Lieutenant Michael Crossley (32 Squadron) receives his DFC, whilst Pilot Officers V G Daw and D H Grice (32 Squadron), D W A Stones and Sergeants A W Whitby and H Cartwright (79 Squadron) await their awards. Squadron Leader Moody, Station Adjutant looks on, with Flight Lieutenant H A B Russell (left) and Flight Officer Nicholls on parade with 32 Squadron and WAAF personnel.
(Photo: Imperial War Museum)

climb, a higher ceiling, and a faster power-dive – all significant enhancements in measuring their performance against the Me109.

At this time too, despite the pace of operations, the Station was visited by King George VI who came to hold an Investiture to honour those pilots who had distinguished themselves so well in the French and Dunkirk campaigns: DFCs for Michael Crossley, 'Jack' Daw and 'Grubby' Grice of 32 Squadron, and Jimmy Davies and 'Dimsy' Stones of 79 Squadron, and DFMs for Sergeants Whitby and Cartwright of 79 Squadron. The ceremony was all the more poignant for at the end a DFC lay uncollected on the velvet table. 79 Squadron had been on operations that morning and Flight Lieutenant Davies had not returned – his DFC never was presented to him . . .

Pilots on stand-by at squadron dispersal.
(Photo: Mrs M A Andrewes)

An aircraft new to Biggin Hill arrived early in July when 141 Squadron came in from Turnhouse equipped with the sleek Defiant fighters. This machine had no wing-mounted machine guns but was armed instead with a mid-fuselage turret operated by a gunner. At this stage there were only two squadrons with them, the other, 264, had operated them with doubtful success over Dunkirk and the machine had shown itself to be vulnerable and perhaps best used at night. However, the pressure of operations and shortage of aircraft dictated that the Defiants be pressed into action – so south they came. The squadron set up shop at Biggin and dispatched pilots and aircraft to West Malling, from where they were to operate out of Hawkinge. Their first action was on 19 July, and it was almost their last. Three of the squadron's twelve aircraft were out of action with engine problems and the remaining nine were patrolling the Channel south of Folkestone when they engaged twenty Messerschmitts. They put up a brave show against the Germans but six of the Defiants failed to return and nine crew were lost. The lack of fixed forward-facing machine guns proved to be a serious fault as far as day-fighter operations for the Defiant were concerned, because the German pilots soon realised that there was a blind spot when they approached from below and to the rear, a fault which made the aircraft easy pickings. As a result of this disaster, 141 Squadron was quickly moved back to the north to retrain and operate exclusively in the night-fighter role.

32 and 610 Squadrons continued with the relentless 'scrambles' and patrols, and the long daylight hours of July saw them down at Hawkinge, often starting before 0500 and finishing at around 2030, and completing three, four, sometimes more, sorties in one day – and always just as many Jerries around at the end of the day as there were at dawn – or so it seemed. As the month progressed the raids grew heavier and more frequent and it was clear that the coastal airfields and radars were the prime targets. Presumably the inland stations were next on the list? Meanwhile, the station score for July was twenty-seven destroyed – a good score after a month of gruelling hard labour, but tiny when compared with what was about to descend.

The Storm Breaks

The first ten days of August were quiet, with most raids outside the Biggin Sector and all of them contained without calling on the Biggin pilots. But then came the Twelfth. This was Hitler's 'Adler Tag', The Battle of Britain was on. The day dawned cold and clear, with 610 on early morning stand-by. 'Scramble' came at 0731, eleven waves of bombers, heavily escorted by fighters, were crossing the Channel. 610, led by Squadron Leader John Ellis, intercepted a group of Dornier 17s off New Romney and broke towards them. A typical melee followed in a protracted action as Me109s raced towards the Spitfires to keep them from the bombers. More squadrons of Me109s arrived, vastly outnumbering the twelve Spitfires, but 610 battled on determined and undaunted, although they could not get at the Dorniers which succeeded in bombing their target – the vital radar station at Dungeness. Suddenly, after twenty minutes or so, all went quiet – the Messerschmitts were retreating back to France to replenish fuel.

Eleven Spitfires returned to Biggin Hill, the twelfth, flown by Flight Lieutenant Smith, had been hit by cannon and he had baled out into the Channel where he was quickly picked up, safe but burned about the face and hands.

Back at Biggin it was still only 0830. The score was two confirmed, six unconfirmed, two probables and one damaged, with all pilots safe and one Spitfire lost – not a bad way to start a major battle that was to set the course of the whole war!

However, in the early morning raids the Germans had taken out five of the

32 Squadron pilots 'scramble' to their Hurricanes in August 1940.
(Photo: Kent Messenger)

Pilots of 610 Squadron await refuelling and re-arming turn round in August 1940. This is a press photograph and background buildings behind the aircraft have been 'painted out' by the censor.
(Photo: Fox Photos)

radars between Dungeness and the Thames and we were temporarily without early warning. At lunchtime 32 Squadron, under Squadron Leader Michael Crossley, was moved down to Hawkinge to be closer on hand. They were 'scrambled' at 1430 but without incident, and then at 1650 they were up again. Two waves of Dorniers with a large escort of fighters were heading for Dover. 32 moved in – in Crossley's words 'What a Blitz – 32 versus the world, it seemed.' As 32's Hurricanes tore into the fighters, the Dorniers moved on to bomb the airfields, Hawkinge, Lympne and Manston – all reduced to shambles – and the Hurricanes airborne with nowhere to land. Carefully, almost daintily, they picked their way through the sea of craters and giant 'mole hills' which half an hour before had been the smooth grass of Hawkinge receding from beneath their wheels. They made it, but as they parked their now unarmed and fuel-less aircraft in came another wave of bombers – by some miracle none of the Hurricanes was damaged, and after refuelling back they flew to Biggin, to be stood down at 2000. So that was 12 August – as Crossley said, 'The first day of the Hun shooting season over the North Downs – not a bad bag – sorties thirty-six, confirmed kills five, unconfirmed sixteen, probables four, and only one Spitfire lost!'

13 August was quiet, but next day 610 encountered Stukas and Me109s off Folkestone and in a brief engagement of a few minutes accounted for six with a further seven probables. Meanwhile, 32 Squadron had ill-luck and lost three of their Hurricanes, although the pilots were not hurt.

15 August saw the Luftwaffe out in strength on what was to be one of the heaviest days of the Battle. The Germans launched some 1,800 aircraft in five huge assaults spread between Kent and Tyneside and lasting throughout the day.

32 and 610 Squadrons were not involved in the early morning raids, which were dealt with by the Hornchurch Sector, and the second onslaught was in 12 Group territory against Tyneside where 79 Squadron, operating out of Acklington, took and destroyed eighteen Huns without loss. The third assault came at 1430 as a two-pronged advance on Harwich/Felixstowe and Dover and Deal. 32 Squadron was 'scrambled' first to Harwich – out of their usual territory – to head off some Me109s fleeing for home. At first they got none but in the chase towards Holland one suddenly turned on Pilot Officer Grice and shot him down in flames, although he escaped unhurt and within a day was back in action. 'Grubby' was immediately avenged after a high speed chase low over the waves by one of 32's Czech pilots whose engine then cut out forcing him to glide back to the coast and force land in a field.

32 then straggled back to Biggin, only to be called out again an hour or so later, this time over Portsmouth where some three hundred Germans were attacking. The final attack that day was closer to home and came at 1830, Dornier 17s and Messerschmitts were approaching London over Kent and 610 Squadron went up to intercept. They met over Sevenoaks. Two Dorniers were downed immediately and

two damaged – the others fled. Another formation, however, got through to its target – Croydon airfield – where the resident squadron, 111, just managed to get airborne and defend its base. 32 Squadron was dispatched post-haste to help and in a fast and furious action they beat off the bombers, taking three Dornier 17s, two Junkers 88s, and four Me109s. So ended one of the heaviest days – a day in which Fighter Command put up one hundred and fifty aircraft in actions which involved all twenty-one of 11 Group's squadrons. The early score was assessed as one hundred and eighty-two enemy destroyed for the loss of thirty-four, but the sheer scale and confusion of the fighting led naturally to over-estimation, for the true figure was later assessed as seventy-six enemy killed – still a pretty good day!

More raids followed on Friday, 16 August, involving 32 in four patrols in which they claimed nine more destroyed, whilst 610 got three probables. The Saturday was quiet but this proved to be another lull before yet another storm for Sunday 18 August was to be the hardest day of all.

The First Bombing Raid

It was midday, and several large formations were well inland approaching London from the south. They hit Kenley first, in two waves one high and one at low-level. 64 and 111 Squadrons rose to defend Kenley, 32 and 610 being held at Biggin Hill, but Kenley was almost wiped out – hangars destroyed, runways and airfield torn up, workshops and domestic buildings smashed and aircraft left burning on the ground. The raid moved on to Croydon and then to West Malling. Was Biggin Hill to be bombed? The answer soon came as fifty-plus Germans were plotted over Tonbridge heading north-west – it was Biggin's turn now. 32 Squadron was 'scrambled' and climbed desperately for height – then 610's Spitfires joined them. The sirens went and the Tannoys warned everyone to take cover. The first bombs fell to the east, in the woods between the airfield and Downe, and caused little damage, but then the sticks rained down. The MT sheds were hit and a Bofors anti-aircraft gun was taken out, one of its crew killed and others wounded. The airfield was raked with bombs, huge plumes of earth were bursting skywards – 32 and 610 ploughed into the raiders in a furious defence of the *Bump*, but they were outnumbered five to one by a mix of Dornier 17s, Junkers 88s, and Messerschmitt 110s. Nevertheless, they broke up the formations and caused the raid to be shortened to no more than ten minutes, after which the bombers turned and fled in disarray, still under chase.

Thanks to the valiant action of 32 and 610 Squadrons, Biggin Hill had come through its first raid quite lightly by comparison with Croydon and Kenley and with the coastal airfields. Some five hundred bombs had fallen, one hundred of them delayed action, but the only buildings damaged were the MT sheds, and no aircraft or vehicles had been hit. Two people were killed and three wounded. It was after this raid that the heroic action of Sergeant Elizabeth Mortimer, a 28-year old

WAAF telephonist, was recognised as the first of three examples by WAAF personnel which were to earn the award of three Military Medals (MMs) to Biggin Hill girls – three out of a total only six MMs awarded to WAAF personnel during the entire war.

Sergeant Mortimer was located at her switchboard in the armoury at the start of the raid, and although surrounded by tons of high explosive she remained at her post in the thick of the bombing, calmly relaying vital messages around the various defence posts as though it was the most natural situation in the world, and as the all clear sounded she took a jeep and went out onto the bomb-cratered and shattered airfield, marking out the unexploded bombs with tall red-flagged spikes.

A Dornier 17 shot down in a field off Blackness Lane, Keston during the first Luftwaffe attack on 18 August 1940. Attributed to Home Guard rifle fire, this kill also showed signs of hits from Bofors shells and machine gun bullets.
(Photo: Imperial War Museum)

As the bombers retreated the station emerged from its various dug-outs, shook off the dust and set to straightening things up. It was all hands to the pumps – Air Force, Army, officers, airmen, WAAFs – everyone moving mounds and filling craters to make the airfield 'flyable'.

And later that day 32 Squadron was airborne again, attacking Germans over the Thames Estuary. Their 'bag' was five more Me109s without loss, but on the return Michael Crossley had to bale out with a glycol leak in his engine – he parachuted into fields near Gillingham to be met by a pitch-fork party!

Bad weather gave respite for the next few days, enabling the weary pilots to replenish and relax, and giving the desperately stretched engineers – the fitters, mechanics and armourers – a chance to catch up on seriously needed repairs and

The Station under German bombing attack one evening in late August 1940. This is a Luftwaffe photograph taken by one of the raiding bombers and found in German archives after the war. Bursting bombs are shown within the dotted lines (1) and the airfield camouflage can be discerned. Other buildings have been numbered by the Luftwaffe and their presumed identity shown in the key. This was remarkably accurate except that the Officers' Mess (4) was shown as store houses.
(Photo: Ministry of Defence)

maintenance and patch up the remaining aircraft into at least a semblance of airworthy serviceability.

German raids restarted on 24 August, but this time the airfields of Essex were bombed and Biggin was not involved, although 610 and 32 Squadrons were back in action on this day, fighting off raids over the Channel.

Replacements

The pace of the battle was by now having a serious effect on men and machines – particularly on 32 Squadron which had been in constant action since Dunkirk – and on 25 August the squadron was moved out to rest in the 'quiet' 12 Group northern territory at Acklington. 79 Squadron, 32's 'child' of the expansion before the war, moved back in to replace them. For 32, it was 'goodbye to the *Bump*' after more than nine years service there, and for 79 it was a case of renewing acquaintances. It was now late August and still the Air Force was holding out; every day counted more desperately than ever, for every day permitted some replenishment of our dwindling force.

More Raids

The Germans now seemed to accept that they would not conquer the RAF in the air. Despite superior numbers, the Me109 and its German pilot was not a match for its British counterpart. Thus, the Germans changed their strategy, and now directed their attention wholly to the airfields – in this way they hoped to force the RAF into submission. The next raid came on Friday 30 August, just before noon – a force of some one hundred and fifty bombers with escorts was approaching London intent on wiping out the airfields. 79 and 610 Squadrons were airborne in good time and tore into the raiders at some height, making them break off course and drop their bombs wide – mainly to the west and north where they caused some damage to Biggin village and caused civilian casualties. Later that day, however, at around 1800, the Luftwaffe returned and this time the station was taken unawares. Despite the lack of warning, a group of airmen managed to dive into a shelter whilst others ran for whatever cover was to hand. The shelter received a direct hit, leaving nothing but rubble, mangled bodies and tattered remnants of uniform. 610 was already engaged on a distant sortie, but 79 managed to get six Hurricanes airborne – although there were only nine Junkers making this attack, through their surprise arrival they caused havoc – a Belfast hangar was hit, barrack blocks, workshops, messes and the WAAF block were badly damaged, and much of the transport, although dispersed, was wiped out. Two aircraft were destroyed on the ground and mains services were cut.

A NAAFI girl was killed as she ran to shelter, and one bomb fell very close to the WAAF shelter – almost a direct hit – the walls collapsed and the occupants were buried, entombed. On the all clear everyone tore frantically at the mangled earth to

Left. The WAAF air raid shelter after an almost-direct hit. One girl died here and many were injured.
(Photo: Wing Commander B C Curwood)

Below. Three heroines walk past the bomb-shattered buildings. The three girls who received Military Medals in recognition of their heroism and devotion to duty during the bombing raids. Left to right: Sergeant Elizabeth Mortimer, Corporal (Assistant Section Officer at time of photograph) Elspeth Henderson, and Sergeant Helen Turner. Flight Sergeant Turner died in 1947 as a result of burns and injuries she received in the bombing. In July 1974, Mrs A A McWatt Green (formerly Corporal Henderson) and Miss Elizabeth Mortimer returned to the Station from their homes in Edinburgh and Stowmarket to attend a ceremony at which three of the station married quarters roads were named *Turner Road*, *Henderson Road* and *Mortimer Road* as a lasting tribute to the three girls.
(Photo: Ministry of Defence)

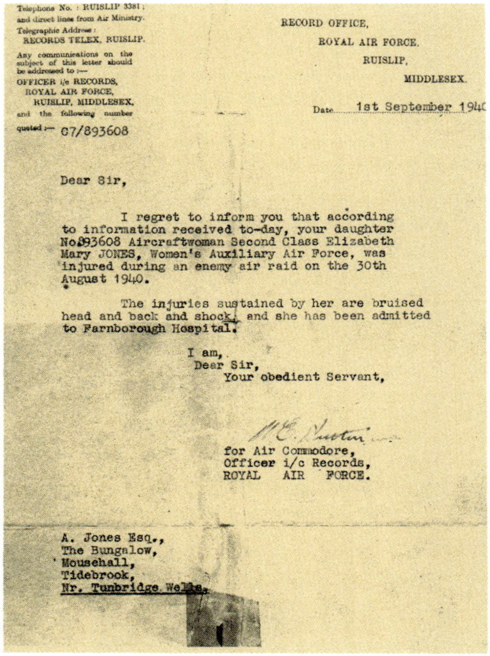

Telephone No. : RUISLIP 3381 ;
and direct lines from Air Ministry.
Telegraphic Address :
RECORDS TELEX, RUISLIP.

Any communications on the subject of this letter should be addressed to :—
OFFICER i/c RECORDS,
ROYAL AIR FORCE,
RUISLIP, MIDDLESEX,
and the following number quoted :— G7/893608

RECORD OFFICE,
ROYAL AIR FORCE,
RUISLIP,
MIDDLESEX.

Date 1st September 1940

Dear Sir,

I regret to inform you that according to information received to-day, your daughter No.893608 Aircraftwoman Second Class Elizabeth Mary JONES, Women's Auxiliary Air Force, was injured during an enemy air raid on the 30th August 1940.

The injuries sustained by her are bruised head and back and shock, and she has been admitted to Farnborough Hospital.

I am,
Dear Sir,
Your obedient Servant,

for Air Commodore,
Officer i/c Records,
ROYAL AIR FORCE.

A. Jones Esq.,
The Bungalow,
Mousehall,
Tidebrook,
Nr. Tunbridge Wells.

Fortunately, it was not long before ACW2 Jones was back on duty.
(Photo: Mrs E M Denness)

release the trapped WAAFs; out they were brought, one by one, most of them safe, but all dazed, shocked and badly bruised. Some were more seriously injured, and one was dead.

Meanwhile, the airmen's shelter struck at the beginning of the raid was still a tomb, and personnel worked into the night on volunteer shifts in vain attempts at rescue. Eventually a team of miners was brought in, the job was so large – but there was no hope for those inside. The death toll on that day was over forty, eleven of them civilian employees.

Next day 610 was pulled out – it was now their turn to rest – and up they were sent to Acklington. They were replaced by 72 Squadron, who were scheduled to arrive later in the day; meanwhile, 79 Squadron was temporarily in sole defence of the *Bump*.

Still More Raids

The Luftwaffe came back on schedule at midday, when a small formation of Dornier 17s delivered a high-level raid – 79 Squadron was out in a dog-fight over Dover and Biggin was undefended. Nevertheless, the Ack-Ack guns put up heavy flak, and caused the bombs to fall wide, mainly to the east, although the runways were badly hit – so much so that 79 Squadron aircraft were unable to land on their return, and had to divert to Croydon.

Once again, it was all hands to the picks and spades as the whole station turned out to level the runways so that 72 Squadron could make their much needed arrival on schedule and 79 could return. At 1800 back came the bombers in another raid. 79 and 72 Squadrons were airborne over the Channel and the ten raiding Junkers had a clear run. Again the runways were plastered and this time most of the surviving buildings were hit. Non-essential Operations Room staff were ordered to shelters for the Operations Block was a vulnerable building, located so close to the airfield and of only light construction. A bomb severed the cables and lines connecting the block with 11 Group Operations Centre and the outside world, but Sergeant Helen Turner and Corporal Elspeth Henderson sat doggedly at their switchboards next to and within the Operations Block vainly seeking to establish contact. Such was their determination that they stayed at their posts even after a direct hit on the building. They had to be forcibly removed after the building caught fire and was evacuated, Sergeant Turner having suffered quite serious injury. For their tenacity they were awarded the second and third of Biggin's three Military Medals. Outside was a sorry sight – the whole station was in ruins and much of it on fire, including the armoury, but somehow there were no further casualties.

On Sunday 1 September Group Captain Grice and Section Officer Felicity Hanbury, the Officer in charge of WAAF who had set such a splendid example of calm leadership, were in the small cemetery off Jail Lane attending the funeral of many of those killed; the sirens wailed heralding yet another raid – the fifth in two days – but thankfully it was brief, the funeral continued and there was little further damage to the station.

Bombed Out

On 2, 3 and 5 September this pattern was repeated – light raids with further accumulating damage – and by now there was really nothing of North Camp left intact. Almost all the accommodation was flattened – the station was billeted out at Keston and in Downe, the Operations Room was working from a butcher's shop in the village and workshops were moved to a nearby riding school.

The large gabled hangar was just a shell, but from the air it had the appearance of being serviceable. Group Captain Grice was convinced that whilst buildings still looked intact Jerry would return, so he set about demolishing the remaining hangar with the controlled use of high explosive. The big bang came at 1800 on 5 September – not a raid but the hangar coming down – and sure enough the Luftwaffe did not return!

But the cause of this was Bomber Command. Early in September some of our bombers had struck at Berlin. This action had so infuriated Hitler that he immediately ordered all bombers to concentrate their raids on London. Thus, the struggling airfields were reprieved and the RAF survived – but the Blitz of London had begun.

92 Squadron Arrives

79 Squadron was by now exhausted and was moved out to Pembrey in South Wales for a rest and a change of scene to convoy and coastal patrols. They swapped places with 92 Squadron who came in on 8 September. 92 – the '92nd Foot and Mouth' – had a reputation as a hard hitting, hard living and hard playing band of young men with an apparent disregard for military discipline and an unorthodox approach to their role as fighter pilots. Nevertheless, it was not the time for polite military niceties, and in the prevailing circumstances their approach proved effective – their casual nonchalence belied dedication, and a jealous spirit of high professionalism. Their stay at Biggin was to be protracted and very soon they were to become the best-known squadron in Fighter Command.

92 was raring to go, they had been living quietly in the west since Dunkirk and they needed action. They found it soon enough and were on intruder patrols from the day of their arrival, helping 79 Squadron fight off the raids on London; massed raids in daylight were now the pattern and the prime target was Dockland.

Crescendo

15 September brought a series of massive raids. The Biggin squadrons were on call from early morning – 72 and 92 were 'scrambled' just after eleven and together the two squadrons intercepted massive formations of bombers with a huge fighter

Left. The Salvation Army kept body and soul together. The tea wagon calls at the improvised workshops in the riding school at Keston in the autumn of 1940. Harry Taylor, the driver, is in overalls fourth from the left. Today, forty years later, Harry Taylor is still a Salvationist – as Brigadier Harry Taylor.
(Photo: G W Chambers)

Below. Breakfast at dispersal, taken on the tailplane of a 79 Squadron Hurricane.
(Photo: M H Thornburgh)

Right. Sergeant Mortimer dons a tin hat to help man the Salt Box Hill gun post.
(Photo: Miss Elizabeth Mortimer)

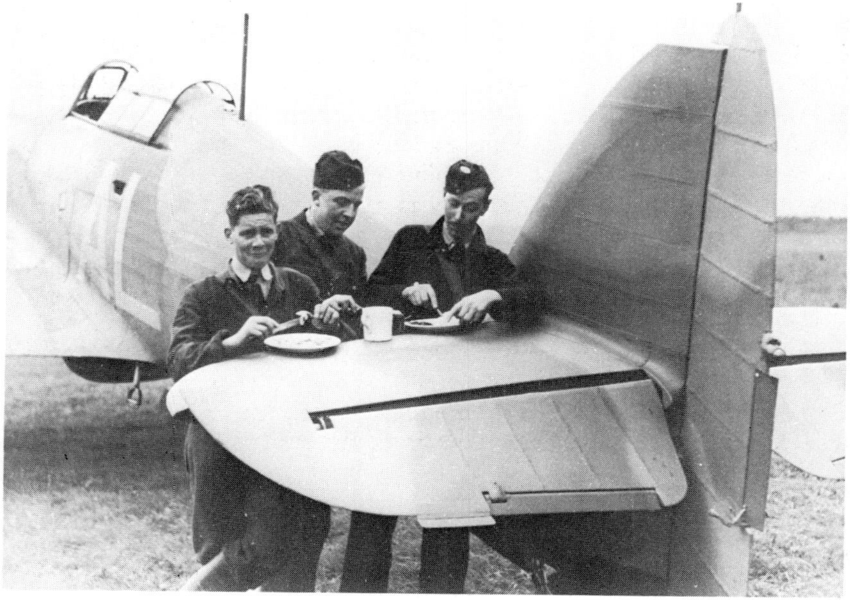

escort over Maidstone. As they approached they were joined by six squadrons of Hurricanes, but even then they were very heavily outnumbered. In they ploughed into a long and confused dog-fight in which they came out on top – nine confirmed and twenty or so damaged, for the loss of one Spitfire whose pilot was only slightly injured.

This day was yet another turning point, for it was a major onslaught by the Luftwaffe and 11 Group's fighters had fought them off. Fifty-six German aircraft were destroyed that day for the loss of twenty-six British. Hitler took stock and finally realised that he had failed in his bid to wipe the RAF from the map. Now he was forced to postpone 'Sea-lion' indefinitely, and indefinitely meant forever. 15 September was recognised as a day of major victory, the most significant turning point in the Battle, and it became designated 'Battle of Britain Day'.

But the fight went on. For a short spell after 'Battle of Britain Day', 141 Squadron returned with its Defiants, now fully converted to the night-fighter role and spent a successful few days on operations against the gradually increasing, but still small-scale, night bomber raids on the capital. The *Bump's* facilities, however, were much too stretched to accommodate a night-fighter squadron alongside 72 and 92, and after only three days 141 moved on to operate from Gatwick.

September continued with busy patrols and gradually the scores mounted. As autumn weather set in with the approach of October the pattern of fighting changed. The bombing attacks moved almost wholly to night raids – the Blitz was on for real – and the daylight hours brought mainly a return to frequent small-scale raids by fighters and fighter-bombers. To combat these rapid and unpredictable raids, regular standing patrols were introduced and these necessarily meant long hours in the air with comparatively little action and small return.

Once more it was wearing and tiring, and the strains began to tell. 72 Squadron, by now depleted to only seven pilots of experience and carrying a lot of 'freshmen', was moved north to Leconfield to brush up with some operational training. Their replacement was 74 Squadron – the famous Tigers under Squadron Leader A G 'Sailor' Malan.

The Tigers were late arrivals in Battle of Britain terms but things were still simmering and they were immediately busy joining 92 in friendly rivalry to form a most effective partnership.

Gradually the pace slackened. 31 October is officially designated as the end of the Battle but there was no clearly defined finish. The fighting continued into November and December but it reduced in intensity as the winter fogs and storms encroached.

The Germans failed, and their failure was a defeat, for it meant that Hitler had lost forever the opportunity that had been fleetingly his. The Battle of Britain had been won, the Battle for Britain was now to continue.

The Station History
(continued)

1941-1980

Recovery

Slowly, very slowly, the station regained the physical semblance of normality, but North Camp was virtually in ruins and for some months the administrative sections operated from requisitioned premises in Keston, with large numbers of personnel billeted out in local houses in Keston, Downe and Cudham. Although repairs were under way by the end of the year, it was a long time before the station really recovered – indeed many of the scars are still borne to this day.

Squadron Leader J A Kent, a tall and imposing young Canadian, took command of 92 Squadron in October 1940 and soon began to use his strong character to discipline his unruly but talented and hardworking team. He certainly made his mark and won something of a trial of strength for, whilst never keeping such a diverse and extrovert group of healthy young men fully in check, he won their respect and was able to mould them into an even more efficient fighting unit.

Flight Lieutenant J C Mungo-Park (in cockpit) and Pilot Officer H M Stephen of 74 Squadron, pictured after gaining the Station's 600th kill in November 1940.
(Photo: Imperial War Museum)

600 Kills

66 Squadron arrived from West Malling in November, having just re-equipped with Spitfire IIas, and by the end of this month the pace of operations was very much reduced by winter weather. Even so, on the 29th the tally of enemy kills stood at 599 – well clear of any other Fighter Command station. The 500 mark had passed unnoticed in the thick of the Battle sometime back in September, but now there was time to make an occasion of the 600th. Bets were taken and many books operated to create a fund for the lucky pilot. There was a buzz of excitement about the station.

The last day of November was fog-bound, grey and dull – all aircraft were grounded – but that magical six-ton figure proved too much for two of the pilots of 74 Squadron, Mungo-Park and H M Stephen. They jointly took off on a voluntary mission to seek out and bring down that elusive 600th. They were virtually the only two aircraft in the whole of Fighter Command airborne that day, and the numbers of Luftwaffe aloft must have been equally minimal. But, despite the weather, they encountered a patrol of Messerschmitts over the Channel off Deal. Together, they picked on a straggler, homed in and shot him down over the coast near Dungeness. Biggin Hill had reached 600.

Another important development at this time was the introduction of the 20mm cannon as main armament in the Spitfire. 92 Squadron began to receive the new Spitfire Vbs early in December, and these aircraft had two cannon and four Browning machine guns – thereby introducing a formidable new fighting machine which packed a very powerful punch.

August 1941. Two of 72 Squadron's Spitfires wait to start up whilst a third taxis to the runway. *(Photo: Imperial War Museum)*

Group Captain A G 'Sailor' Malan, DSO and Bar, DFC and Bar, was probably the outstanding fighter pilot of the war with 35 kills and 11 probables; he was never shot down and never even had to bale out. He was part of the fabric of Biggin Hill; at the time of this sketch by Cuthbert Orde he was CO of the crack 74 Tiger Squadron. 'Sailor' stayed on as a Wing Commander to be the first commander of the Biggin Wing and then in 1943 he returned as a Group Captain to command the station. After the war, he returned to his native South Africa, where he died in 1963.

The end of 1940 also brought the award of many well-deserved decorations for Biggin Hill pilots: DSOs for 'Sailor' Malan and H M Stephen, a DFC for John Mungo-Park and nine others, and many DFMs. Such awards were not given lightly – but represented recognition of exceptional bravery, sacrifice and sustained devotion to duty. Group Captain Grice, the ever-popular and much respected station commander who had led the station through the thick of the Battle of Britain, moved on in December 1940, handing over command to Group Captain F 'Mongoose' Soden.

February 1941 saw 66 Squadron move out to Exeter, their replacement being 609 (West Riding) Auxiliary Squadron from Warmwell. By this stage of the war – like most of the '600' Squadrons – 609 was Auxiliary only in name, and now had but a tenuous connection with its Yorkshire origins. Many changes then followed in quick succession; 74 Squadron moved on to Manston for a few months and then in April to Gravesend (back in the Biggin Hill Sector). Meanwhile, 264, with Defiant NFIs, spent January to April operating from the *Bump* in successful night-fighter attacks on the Luftwaffe bombers engaged in blitzing London.

We Take the Offensive

As the spring of 1941 approached, Biggin Hill turned to offensive operations over France. The technique of wing sweeps was developed with Malan commanding 92, 609 and 74 Squadrons working in unison, and in co-operation with the Tangmere Wing, a similar formation operating under the command of Douglas Bader. Offensive operations rapidly increased in scope and intensity, and the era of 'Rodeos', 'Rangers', 'Rhubarbs', 'Roadsteads' and 'Circuses' etc was born. To explain:

Rodeos were fighter sweeps over enemy territory.

Rangers were large-scale fighter sweeps (Wing sweeps), aimed at drawing enemy fighters into combat so that they could be picked off and the Luftwaffe gradually worn down.

Rhubarbs were small-scale (two or three aircraft) fighter attacks on opportunity targets.

Roadsteads were attacks on ships at sea or in harbour, by fighters, fighter-bombers, or bombers escorted by fighters.

Circuses were small to medium-scale bombing attacks by bombers escorted by fighters on short-range and fringe targets, aimed at drawing and engaging enemy fighters.

These operations built up in various forms and combinations into 1941, and by the middle of the year, when Group Captain Philip Barwell had taken command from 'Mongoose' Soden, the Biggin squadrons were very active and experienced in the aggressive role, the long-term aim being to wear out the Luftwaffe through a process of attrition.

Scores were high, fifty Germans destroyed in June and July, and a further

1941. Flying Officer Ziegler, 609's Intelligence Officer, interrogates a squadron pilot after combat, whilst his Spitfire is rearmed and refuelled.
(Photo: Central Press)

thirty-five probables. 92 reached their one hundred and fiftieth Hun, and at last the boot was on the other foot, for the battle was over enemy territory and it was the Luftwaffe's turn to be on the defensive and constant alert against the invader. But all this was not without casualties, and with the scene of battle now over France and the Channel many aircraft were lost, especially at sea, and more than a dozen Biggin pilots owed their lives to the lumbering Lysander patrols of the Air-Sea Rescue Service and to the high-speed launches of the RAF Marine Branch. Even so, Squadron Leader Mungo-Park, now CO of the Tiger Squadron, was killed in a raid on St Omer, and sadly he was but one of so many.

July brought more movements – 609 Squadron moved out to Gravesend to swap places with 72 who replaced them on the *Bump*. 'Sailor' Malan handed over to Mike Robinson as Wing Commander Flying and went off to a liaison tour in the States, and the summer continued with more wing offensive operations. In October 401 (Canadian) Squadron arrived from Digby with Spitfire Vbs to join 609 who were now back, having again swapped Gravesend with 72. 609 then departed north to Digby in November as 124 Squadron moved in from Castletown to join 401 as the only operational squadrons in residence for the winter of 1941-42.

Scharnhorst Escapes

That winter brought only limited action, but excitement came early in February when a chance patrol from Kenley spotted the German battle-cruisers *Scharnhorst* and *Gneisenau*, and the cruiser *Prinz Eugen*, out of Brest and slipping back to Germany up the Channel towards Dover under cover of fog and heavy cloud, and under the very noses of the Navy and Coastal Command. With the weather so bad, Biggin had not expected any flying, and when the alarm was raised the station was caught napping. 72 Squadron, however, at Gravesend, managed to 'scramble' and rendezvous with six Fleet Air Arm Swordfish out of Manston, and escort them out into the Channel to track down and attack the battle fleet with its heavy escort of fighters, frigates and destroyers. The Swordfish, each armed with a single torpedo, and led by Lieutenant Commander Esmonde, pressed home their attack against overwhelming defensive fire from the German ships. All six Swordfish were lost, but two of them got their torpedoes through and the action brought Esmonde a posthumous VC.

The Biggin Hill squadrons arrived on the scene to join the chaotic melee as Beauforts armed with torpedoes, and naval MTBs moved in under heavy attack from defending guns and fighters. 124 and 401 Squadrons did what they could to contain the fighter escorts, but as RAF Wellington bombers moved in, the Biggin units were forced to withdraw by the atrocious weather. *Scharnhorst* and *Gneisenau* managed to run the gauntlet and reach their homeland but it was later learned that they were both critically damaged, and were virtually out of action after hits by many of the aerial mines dropped from the Wellingtons.

Bruneval Escort

Bob Stanford-Tuck had taken over as Wing Commander Flying in December 1941 and the continued winter weather inactivity caused him mounting frustration for he was a man of airborne action and had not been born to sit around. A clear day dawned on 28 January 1942, so Tuck and a colleague found time and opportunity for a Rhubarb over France. They swept in fast and low, inland from Le Touquet, and successfully attacked a variety of tempting targets; their presence, however, soon drew a mounting barrage of flak and Tuck's Spitfire was hit, forcing him to make an emergency landing in a heaven-sent field. He was not injured and he thus became a prisoner – once more Biggin Hill needed a new Wing Commander Flying.

Wing Commander Masterman filled the gap and the shock of Tuck's capture was dispelled by an unusual diversion on 28 February. 124 and 401 Squadrons were put on early morning alert and were 'scrambled' as escort to two destroyers returning up-Channel from France. Twenty-four Spitfires to escort two small destroyers was unheard of, and the station wondered what on earth was afoot. Unbeknown to Biggin, these two ships were carrying the sappers and the paratroopers who had just completed the outrageously daring commando raid on Bruneval, and were bringing back with them, serviceable and intact, the 'Wurzburg' top secret German anti-aircraft tracking and guidance radar.

As the spring set in so the pace of operations speeded up. In March, 72 Squadron did yet another repeat of its swapping act at Gravesend, and moved back to the *Bump* as 401 replaced them by the Thames. April saw the departure of the Army when they handed over airfield defence to the newly-formed RAF Regiment, and finally severed twenty-five years continuous presence on the station.

HRH The Duke of Kent is welcomed to the Station by Group Captain Barwell on a visit early in 1942. Later that summer both were killed in the air.
(Photo: Imperial War Museum)

The Eagles Arrive

As offensive sweeps and raids over France continued into May, a new squadron came in to partner 72. 133, the third of the three Eagle Squadrons, comprising volunteer Americans flying as RAF pilots before the United States joined the war, came to Biggin to be blooded into its first action. They soon settled and worked well alongside the seasoned 72, playing a very full and highly professional part in all the operations that were going.

The first day of July 1942 brought a tragedy to the station. For some time a high-flying German reconnaissance aircraft had been intruding into the Sector airspace during the early evenings. Biggin had taken delivery of a few Mk VI Spitfires which had been allocated to 72 Squadron. These aircraft were modified for high altitude work, having a pressurised cabin and specially extended wing tips – giving them an odd appearance. However, they were just the thing to tackle the high-flying intruder, and the Station Commander, Dickie Barwell, went up with Squadron Leader Oxspring to stooge at altitude and intercept the Luftwaffe 'spy'. Once on station, Barwell suffered complete R/T failure, and was not aware that two other Spitfires had left Tangmere on the same mission. The four aircraft met over the Channel, and the Tangmere pilots momentarily mistook Barwell's strange Spitfire for the intruder. They attacked and Barwell crashed off Eastbourne. He was killed.

The station was numbed and took some time to resettle under the command of Group Captain Hallings-Pott. 72 and 133 briefly moved out to Lympne for a few days and were replaced by 19 and 234 Squadrons. The Lympne detachment was shrouded in mystery, but was later resolved as readiness for the assault raids on Dieppe, which had been planned for July but were postponed when a group of the assault craft, fully laden with Canadian troops, was shot up and bombed by the Luftwaffe in the Solent.

Biggin then received the high-performance Spitfire IX as replacement for the V, and more squadron changes quickly followed. 72 finally left Kent and went north to Morpeth. It was replaced by 401 – the Canadians – back again from Gravesend. The summer of 1942 was dubbed the 'American Summer', for the resident 133 Eagles and the Canadians of 401 were to play host to newly-formed USAAF units which had crossed the Atlantic (without their aircraft – which were following on ships) as the vanguard of the USAAF in Britain, as America joined the war. These units formed the 307th Pursuit (Fighter) Group, and they flew in borrowed Spitfires to gain operational experience.

Dieppe

August brought the resurrection of Operation Jubilee – the Dieppe Raid – and off again went the squadrons to Lympne, this time for real. 222 and 602 Squadrons came in to fill the gap and swell the wing strength, and the USAAF 307th Pursuit were delighted to be invited to take part. Some 7,000 troops, mostly Canadian, were landed at Dieppe at dawn on 19 August. The raid was planned as something of a practice or exercise for bigger things to come, and the air operations put the embryo Second Tactical Air Force Units through some testing paces. Fighter Command fielded a total of fifty-six fighter squadrons, and the role of the four-squadron Biggin Hill Wing – comprising 133, 222, 401 and 602 plus the 307th Americans, was twofold: to help maintain complete fighter air cover over the invading troops as part of a through-the-day shift system, and to kill as many of the Luftwaffe as possible.

The Wing made four sorties that day: the first at 0600 was uneventful, but on return at 1000, kills ran to good firm figures with probables well into doubles. The midday patrol produced only few results and a final sortie in the early evening was wholly nugatory.

Overall, the raid was expensive. The invading troops were ashore for nine hours, a lot of damage was done to military and civil installations but 3,500 troops were lost – in just one day – and the total aircraft score was low – only forty-eight destroyed and twenty-four probables; against RAF losses of one hunderd and six.

Pilots and groundcrew stand atop a dispersal shelter and crane skywards to watch Spitfires returning from a sweep in 1942. Pilots from 133 Eagle Squadron are included in this group. *(Photo: Imperial War Museum)*

Even so, the lessons of the day were invaluable in the later planning and conduct of the North African, Italian and D-Day landings, and those 3,500 lives lost undoubtedly saved many more thousands in these later campaigns.

133 is Lost

The return to normal offensive operations was something of an anti-climax after the intensity of Dieppe, and once more things were quiet. In September, the decision was taken to disband the three Eagle Squadrons and absorb them into the USAAF which by now was building up its forces in the UK. 133's final RAF operation was on 26 September and was an unparalleled disaster – a tragic end to a short but very distinguished career. The Biggin Wing was escorting American Fortresses in a bombing raid on Brest. 133 Squadron was detailed as the distant escort and went ahead to refuel at Bolt Head in Cornwall, this being necessary to ensure maximum time over the target area. The Meterological Officer at Bolt made a serious and costly error in giving them wind forecast, warning them of strong southerlies instead of northerlies – he was 180° adrift. The squadron made rendezvous with the Fortresses and together they headed south through heavy cloud. Conditions were bad and 133 had to abandon the escort and return to Cornwall short of fuel; when they came down through the cloud they found to their horror that they were still over Northern France – not Cornwall – and their tanks were almost dry. Eleven of them had to make forced landings whilst the twelfth managed to reach England only to crash into the cliffs of the Lizard. The whole squadron was wiped out without a shot fired.

Three days later in a ceremony at Debden, the three Eagle Squadrons, 71, 121 and the remnants of 133, were transferred to the USAAF, having set a proud record in a short few months of devoted and gallant flying.

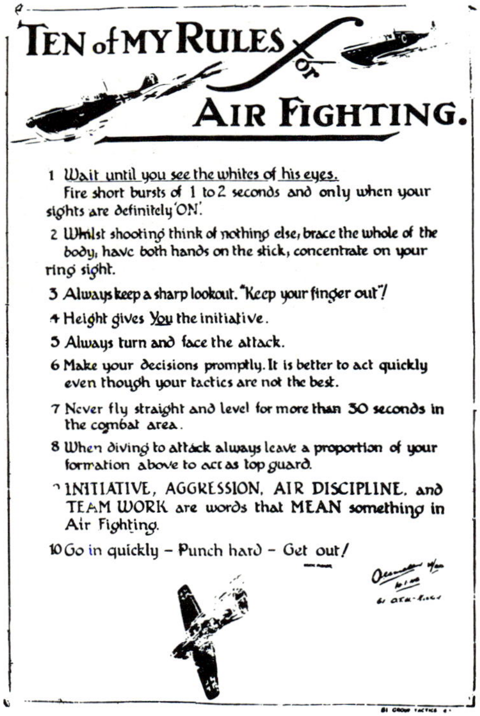

The Free French

340 (Free French) Squadron, the famous 'Groupe Ile-de-France', came from Hornchurch in September 1942 together with 611 (County of Lancashire) Auxiliary Squadron from Redhill. By now preparations were in hand for the invasion of North Africa, and Fighter Command offensive operations were subdued. As winter progressed, the Biggin squadrons settled in to a routine of long and uneventful hours spent on bomber escorts.

The new year brought the return of 'Sailor' Malan, now as Station Commander and in the rank of Group Captain. The dull and routine life of escort duty was sapping morale, and Malan was faced with the task of lifting the spirit of his command. Quickly he instilled his zest and enthusiasm and his competitive quest for the highest professional standards. Out of the cupboard came his 'Ten Commandments' – his rules for air fighting. 'Sailor's' presence was instantly felt and the station was quickly galvanised into a renewed sense of pride and purpose. The onset

Dawn take-off for 340 (Free French) Squadron in January 1943. *(Photo: Imperial War Museum)*

of spring saw a return to Rodeos, Rhubarbs and Circuses whenever weather conditions permitted, but, despite the renewal of high morale, the station went through a period of losses – heavy losses considering the relatively light loading of operations, and strangely it was the more experienced pilots who were becoming casualties.

Wing Commander Alan Deere, the young New Zealander, was now OC Flying, and No.1 Squadron joined 340 and 611 in February 1943, coming with the new Typhoons for a brief five-week stay before moving on to Lympne in March. 340 then went north to Turnhouse and were replaced by the other famous Free French squadron, 341 (Alsace) under Commandant Rene Mouchotte.

The 1000th Hun

Mouchotte was a quiet but determined and highly professional aviator. Only the best was acceptable to him and his squadron had already set a remarkable record in its few short weeks of life. The Frenchman was wholly engrossed in flying, and he permitted his squadron to take part in operations only after the most intensive training had enabled them to master their leader's unusual and unorthodox tactics in the air. Mouchotte's total dedication was remarkably successful, and as 1943 progressed the once elusive Sector total of one thousand enemy kills was looming close. On 7 May it stood at nine hundred and ninety-five, and on 14 May at nine hundred and ninety-eight.

Commandant Rene Mouchotte and Squadron Leader Jack Charles share the honour of making the Station's 1,000th kill in May 1943. Mouchotte was killed in action three months later.
(Photo: Ministry of Defence)

15 May brought a late afternoon bomber escort over Caen, and the game was on. The Operations Room was full to the brim, and the excitement was intense. Both 341 and 611 were aloft, and the target area had been reached without incident. As the first bombs were released a pack of FW190s climbed rapidly towards the raiders, 611 went in first – 341 giving them top cover – and Squadron Leader Jack Charles quickly took a victim – the nine hundred and ninety-ninth – and then homed-in on another. At precisely this moment Mouchotte had a 190 squarely in his sights and let fly on all guns. Two more German aircraft were crashing in flames but which of these was No.1000? There was of course no way to tell and the honours were shared – Mouchotte and Charles.

The Station, the Sector, 11 Group, Fighter Command and the National Press went mad – photographs, statements, interviews, parties and the 'ticker-tape' of a wartime 'Good News' item. A vast celebration was organised at the Grosvenor House in London, and was attended by the top brass of Fighter Command, and, by hook or by crook, most of the RAF pilots (plus a host of others) in Kent at the time. The revelries were so extensive that final settlement of the bill was achieved by a levy on all Fighter Command pilots – whether or not they attended!

The party over, the summer of 1943 saw continuous raids on the Luftwaffe in France – it was rather like the Battle of Britain in reverse – and although the opposition was not as strong, the hours were as long and the strain almost as hard.

. . . And the whole of 611 (left) and 341 Squadrons and their groundcrews join in the 1,000th Hun festivities.
(Photo: Imperial War Museum)

The squadrons had to be pulled out to rest and recuperate. 611 went on to Matlaske early in July and were replaced by 485 (New Zealand) Squadron. But Mouchotte and his 'Groupe Alsace' refused to rest – the action was all over their homeland and they fought steadfastly on and on. Mouchotte had personally completed in excess of one hundred and forty operational sweeps and despite a growing weariness he persevered. Later in July he led the Wing in a most successful action over Triqueville in which he took 341 and 485 through a surprise attack by three squadrons of FWs, climbed up into them and in the space of eight minutes they killed nine of them without any loss. One month later he died in a heavy FW attack whilst escorting Fortresses over St Omer. He was sadly mourned, but as always in wartime, operations continued, with his replacement, Duperier, soon in post.

The 1,000th Hun was, however, the inspiration for a Station Memorial Chapel, and since the war Mouchotte's name has lived on as *Mouchotte Close* – a road in the station officers' married quarters site.

Invasion Plans

The autumn of 1943 brought further significant changes in the conduct of the war. On all fronts the Allies were now on the offensive, the success of the North African campaign had led on to Sicily and into Italy, the tide had turned in our favour in the Atlantic, and the air campaign against the Luftwaffe, over France, Belgium, Holland and Germany itself, had achieved the desired effect of wearing them out and running them down. The Allied air forces were finding airborne opposition increasingly hard to find.

Final plans were now being laid for a massive Allied invasion of Europe, and new concepts of warfare were being developed. North Africa had proved that a self-contained mobile tactical air force was essential to any strategic offensive operation. The concept had been brought to Europe as the invasion plans were formulated. Fighter Command had been reorganised in the early summer to include a complete tactical air force (the Second Tactical Air Force) within its make-up. Other squadrons had been earmarked to continue in the fighter-defensive role, but the Command would retain its present title and function until sometime nearer the planned invasion. It would then divide; the defensive squadrons becoming Air Defence of Great Britain, and the tactical squadrons moving with 2 TAF to become part of the joint Anglo-American Allied Expeditionary Air Force. Meanwhile, the tactical squadrons had to train in their new role.

In October, 341 and 485 Squadrons moved on to train-up as part of 2 TAF, 341 to Merston and 485 to Hornchurch. Biggin Hill was now alllocated to 2 TAF and within the mobile concept it became '126 Airfield', comprising three Canadian squadrons, 401, 411 and 412, all arriving early in October and equipped with Spitfire Vs. This 'Airfield' or 'Wing', as the term was later to become, had spent the summer months of 1943 under canvas in rural Kent at the Advanced Landing Ground near Staplehurst, living rough as they trained in tactical support operations. They quickly re-equipped with Spitfire IXs and commenced bomber escort duties, whilst still keeping 'in touch' with their ground attack role. The shortish range of the Spitfire meant that this trio of squadrons was used as withdrawal escort – meeting the massive formations of American Fortresses on their return from raids over Germany, and shepherding them safely home. Even on this task, the Spitfire was often operating at the limit of its endurance and there were many diversions and crash landings, and even lost aircraft, caused by lack of fuel.

In April 1944, as the planned invasion deadline approached, 126 Wing moved on to Tangmere to be nearer their assigned patrol areas, and they were not replaced. Biggin became a Base Defence Station within 85 Group of 2 TAF, and the *Bump* was without aircraft for most of the summer of 1944. However, its sector-member stations to the south, particularly Newchurch and Brenzett in the Romney Marsh and Lympne on the edge of the Weald, were very much involved in the Normandy landings and the aftermath of the invasion.

The V1 Attacks

By late 1943, intelligence reports and photographic reconnaissance had revealed the existence of rocket launching sites along the coast of Northern France – these were the V1 flying bombs – the 'Doodle Bugs' of 1944. Systematic raids by Bomber Command and by 2 TAF had destroyed most of the coastal sites, but there was a menacing arc of well-hidden and camouflaged sites, inland and safely behind German lines. Their target was London and they would soon be ready to launch.

Defences against the Doodle Bug were quickly devised, the first line being to bomb the sites, and the second being fighter attack over the Channel – these became known as 'Anti-Diver' patrols and were mounted from the coastal locations – Brenzett, Newchurch, Lympne, Hawkinge, etc. for reasons of speed and proximity. Further defence lay in Ack-Ack attack from the coast and in Fighter and Ack-Ack attack over land (ie, mainly Kent) on those that had got through. The last line was a balloon barrage in the style of the early years of the war. The North Downs presented a natural site for such a barrage, the high ground forming a continuous defensive ridge around the south of the capital. Sitting atop the Downs, Biggin was the natural choice for the balloon control centre, and in the spring of 1944, 5344 Airfield Construction Wing moved in with a balloon unit to prepare the sites and access roads for over three hundred balloons.

The first V1 fell near Gravesend on 13 June, and the next day saw 945, 946 and 947 Balloon Squadrons in residence, shortly to be joined by 953 and 958. Within a few days these units had put up a barrage of almost three hundred and fifty balloons at 4,500 feet in a huge line stretching from Tatsfield, over Westerham and Brasted to Ide Hill.

The V1 menace continued after D-Day into early September – perhaps by way of German retaliation as well as a last desperate attempt to mount some sort of offensive action – and Kent became known as 'Bomb Alley'. In all, some 8,500 bombs were launched against London, some 2,500 got through, with a further 2,000 plus falling in Kent. Thus, about half were destroyed – and two hundred and seventy-nine of them by the balloon barrage. The attacks from the south ceased early in September as the Allied armies sacked the launching sites in their advance through France. The balloons were taken down and moved into Essex for a brief period to give protection against V1s launched from aircraft over the North Sea.

A Final Fling

Aircraft returned to Biggin in October, when 322 (Dutch) Squadron came with Spitfire IXs for a spell of bomber escort duty and armament practice, before moving on late in December to operate from Woensdtrecht – their home soil once again! 340 and 345 Squadrons – the latter fresh from Arnhem escort and patrol duty – arrived in November to spend the winter of 1944 using their Spitfires on long-range daylight bombing escorts over the Ruhr and the very heart of Germany. They were joined in these operations by 154 Squadron which had disbanded in Marseilles in October, equipped at first with Spitfire VIIs but very quickly to convert to Mustangs. The long range of the Mustang enabled 154 to operate effectively from UK unitl the very end of the war, and it continued with busy escort duty over the Ruhr and the Rhine, until moving to Hunsden late in March 1945 to disband.

The War is Ended

These were to be Biggin Hill's final wartime operational duties, but the Allied

Victory in Europe brought an urgent need for transport bases close to the Continent. The station thus became a Transport Command unit and housed 168 (Heavy Transport) Squadron RCAF. This unit flew bulk mail across the Atlantic from Canada in transport Fortresses, and then redistributed it from Biggin, using Dakotas to fly it on to Canadian troops throughout Europe. They stayed until the summer of 1946, when Biggin was transferred to Reserve Command.

Auxiliary Station

The Auxiliary Air Force was reformed on 31 July 1946, and on this day two Auxiliary Squadrons, 600 (City of London) and 615 (County of Surrey), moved in to make the station their post-war home. 600 came with Spitfire FR14es, and 615 had a mix of F14s and F21s. Both squadrons remained for the rest of their careers re-equipping with Meteor F4s in 1950, and F8s in 1951. They were the 'Weekend Air Force' but they reached a very high degree of efficiency, and in November 1949 they reverted to Fighter Command, although not as front-line units.

Front Line Again

The Auxiliaries 'held the *Bump*' until March 1951, when 41 Squadron moved from Church Fenton with Meteor 8s. Biggin Hill was thus once more a front-line fighter station for what were to be the last few years of its flying life. 41 re-equipped with the Hawker Hunter F5 in the autumn of 1955, and it is fitting that this fine aircraft, a direct descendant of the immortal Hawker Hurricane, was to be the last operational fighter to fly from the *Bump*.

The rapid advances in aviation technology rendered the continuance of an airborne auxiliary air force both impracticable and impossibly expensive. Thus, 600 and 615 Squadrons were disbanded in March 1957, when the reluctant decision was taken to reduce and reconstitute the Auxiliary Air Force. 41 Squadron flew on alone for a year, but modern warfare and the threat against Europe and the UK had changed. There was no longer any need to locate manned fighters in the crowded air space of South-East England, and so on 16 January 1958, 41 Squadron was disbanded. It reformed the same day at Coltishall, equipped with Javelins and renumbering as 141 – appropriately the number of the first operational squadron ever to be based at Biggin Hill.

Biggin Hill was to fly no more, and Flying Training Command took administrative control of the station. The airfield was relinquished to the Board of Trade and later bought by Bromley Borough Council, who have since operated it as light aircraft and general aviation unit under licence to Surrey Aviation Ltd.

Officer and Aircrew Selection

With the aircraft gone, 1958 was spent under care and maintenance in preparation for the move of the Ground Officer Selection Centre (GOSC) who came from

Groundstaff beside a Dakota of 168 (Heavy Transport) Squadron RCAF in 1945.
(Photo: RAFA Sussexdown)

A Hunter F5 of 41 Squadron parked at dispersal in the autumn of 1955. The 'Seco' buildings in the background were demolished early in 1978.

Candidates undergo leadership selection tests at the Officers and Aircrew Selection Centre.
(Photo: RAF Biggin Hill)

Uxbridge into the North Camp accommodation in April 1959. A major new build then took place next to the Chapel, on the site of the Belfast hangar destroyed by the Luftwaffe in 1940. A purpose-built interview and selection block was put up in preparation for the move of the Aircrew Selection Centre from Hornchurch, and the amalgamation of the two selection units into one. This was completed in the summer of 1962, and the new unit was named the Officers and Aircrew Selection Centre. The Cranwell Selection Board, now the Cadetships Board, joined them in 1964, thus concentrating the whole selection process under one roof.

For many years the operational link was maintained, when each September on Battle of Britain Day the station opened its gates to the public to stage the spectacular and traditional 'At Home' air display. Alas, economies forced this to be discontinued after 1976, although the tradition is carried on, to some extent, through RAF participation in the annual commercial 'Air Fair' held on the civil airfield each spring.

Today, this most famous of all fighter stations continues as an RAF unit. Not a unit with a glamorous role, but a role which is nevertheless vital to the continued well-being of the Air Force, for every aspiring officer or member of aircrew must attend at Biggin Hill for selection assessment. Standards are necessarily high, and of

An aerial view of the Battle of Britain 'At Home' Day, 4 September 1976 – the last such day. The flat-roofed 'E' shaped building at left background is the OASC main block and stands on the site of the large Belfast hangar wrecked in 1940. The Officers' Mess is behind this block and the white on the top of its roof is the balcony – the site of the 1940 gun post. The large hangar to the right is post war. The aircraft are – left to right – Harrier, Hercules, Gazelle helicopter, Jaguar, Gnat, Bassett, and Dominie. This is a detail from a 'crowd-counting' photograph taken by an RAF reconnaissance aircraft at 250 feet.
(Photo: Ministry of Defence)

the many thousands of hopeful young men and women who pass through the Centre each year, a very large proportion must leave with disappointment, for the Station motto – *The Strongest Link* – applies as much today as forty years ago.

Some of The People

The Biggin Hill story and the story of the Battle of Britain have often been told, and they are usually presented as factual and historical accounts of the actions and engagements leading to this vital victory.

But the victory was achieved by people – by a whole team of people, Service and civilian, who worked together with a common aim. To some extent their dedication, example and sacrifices have been officially recorded, but much of the detail of their personal experiences has not. Many of them have a tale to tell . . .

In 1940, Squadron Leader John Worrall was in command of 32 Squadron. He led this unit through Dunkirk and the early stages of the Battle of Britain, and in August he became Senior Controller at Biggin Hill. He remained in the service after the war and went on to become Air Officer Administration, Flying Training Command (1956-58), and Senior Air Staff Officer, Near East Air Force (1960-63), before retiring in 1963 in the rank of Air Vice-Marshal. He writes . . .

During the raids in September many 'shot-down' German aircrew were parachuting into Kent. One landed in a field close to the Westerham-Sevenoaks road and was spotted by the driver and conductor of a local bus. They stopped, and armed with spanners from their toolkit, they went into the field and warily approached the German. On seeing two uniformed men approaching him, the German sprang to attention, gave the Heil Hitler, and in halting English proudly announced 'We have bombed flat Biggin Hill.'

This infuriated the driver, who raised his spanner and dotted the German hard, saying 'That's knocked you flat too!' – Such was the spirit in Kent at the time.

And he adds . . .

Before the RAF Regiment was formed, the Army provided Ack-Ack units and airfield guards and sentries. Biggin Hill was provided with a company of reservists who had not seen active service for years and who, whilst efficient enough at their duties, had little knowledge of the Air Force and its ways and structure. Group Captain Grice was surprised at the friendly attitude of the sentries, who used to nod and smile and greet him in a 'gafferish' way as he went about his duties as Station Commander. He arranged for the Station Warrant Officer to give these friendly elderly soldiers some basic instruction in Air Force ranks and discipline.

There was an immediate and marked improvement, but a few days later on passing one of the sentries, Grice enquired if he was not entitled to a salute. 'Oh no Sir,' came the reply, 'You blokes with shiny peaks on your caps, you's warrant officers!'

Early in August 1940, Michael Crossley, one of 32 Squadron's most distinguished pilots, was promoted to Squadron Leader and took command of the squadron from John Worrall. He led 32 through the remainder of their fighting in the Battle, and at the end of the month took them out to 'rest' at Acklington. For his gallantry, skill and determined leadership, he was awarded the DSO and two DFCs. After the war he left the service to become a farmer, today he lives in the Transvaal. As Squadron Commander he personally kept the Squadron Diary, a document which has been much quoted in subsequent histories and accounts of the Battle. Two verbatim entries are typical of the record and give vivid illustration of the times.

15 August. Thursday

Down to Hawkinge 1pm and from then on had a remarkably blitzy afternoon. Chased something up to Harwich and got mixed up with 109s going home. Got none. They got 'Grubby' Grice instead, who descended into the sea (back a few hours later unhurt). Back to Biggin, refuel, off to Portsmouth and attacked thousands of 88s and 110s. Got three. Refuel again and attack thousands of 17s who were beating up and bombing Croydon. Slapped down seven. 'Polly' Flinders took training flight out and he and Humph slapped down one each. Day's bag twelve.

16 August. Friday

Didn't go down to Hawkinge. Hooray! Called off in A.M. got a bit split up and Red Knight (me) and de Grunne slapped down a 109 each. Just going to lunch when called off again. Saw nothing. Lunch at 3.45pm. Called off again and having got split up while climbing through cloud, seven of us met up with 80+ of them and slapped down eight.

Tried to entertain German prisoner from a 110 in the evening, wouldn't play. Not a nice fellow.

In later correspondence he elaborates on the prisoner:

He told us his name was 'Helmut'. He was a shortish thick-set fellow with a bull-neck and very fair hair, and typically arrogant bearing. The MPs brought him to our dispersal and we showed him a German gas mask, a 109 rudder, and a 110 wing-tip which we think was his, as his machine fell not far from Biggin, and one or two of our lads rushed to the crash to get a piece and prove it was ours. He looked at the trophies and flatly denied that they were even of German origin! We plied him with tea and cake and smoothed his ruffled feathers until the MPs decided it was time for him to go.

I couldn't swear blind that he was 'mine'. I did get a couple of 110s on the 15th and one on the 16th, though whenever I have remembered that evening I have always thought of him as 'mine'.

Well, I do hope some of this may be of use to you. It has given me great pleasure digging back into the old Diary and bringing old faces and incidents into focus. And it all happened close on forty years ago . . .

In a report to HQ Fighter Command, Flight Lieutenant De La Torre, the Unit Intelligence Officer, referred to the sighting of a 'Gaggle' of enemy aircraft. The reference appealed to the sense of humour of the Duty Air Commodore who, after protracted exchanges, sent the following reply via the Intelligence Staff at HQ 11 Group:

I QUAIL at the sight of the POULTRY effort to ROOK you of an excellent colloquialism and I feel that this LARK has been carried too far, and that the gag has by now been completely taken out of GAGGLE. My only GROUSE is that it is not my PIGEON, and consequently I can SWALLOW no more.

(Reported by Flight Lieutenant C E Rogers [Retd], Assisant Intelligence Officer Biggin Hill, August 1940 and Intelligence Staff HQ 11 Group, 1940-41.)

Mr L W Rea of Alfold, Cranleigh, Surrey, enlisted in the RAFVR in September 1939, and was posted to Biggin Hill in June 1940 as a nineteen year old Aircraftman Fitter 2E straight from training at Hednesford. He was allocated to 32 Squadron and he went with them when they were pulled out of the Battle to rest and re-equip at Acklington. Later hed went overseas with 261 Fighter Squadron which flew Hurricanes and Thunderbolts in the Army close support role. He went with 261 through Iraq, Palestine, Egypt, Ceylon, India and Burma, and he was demobbed as a Sergeant early in 1946.

After the war he remained in aviation engineering, serving for 33 years with British Airways in many parts of the world. Now retired, he writes:

I remember arriving at Biggin on a lovely summer's evening and looking down on the Hurricanes parked on the grass by the Hangar. They were immaculate – like beautiful butterflies on a billiard table – but the scene was soon to change.

I remember working with the civilians who were fitting the 'magic' constant speed units and props, and were helping with the training flight, commanded by Flying Officer 'Polly' Flinders.

I remember Czech and Polish pilots, exasperated with the English language and unable to differentiate between throttle and pitch control levers – and some finished up in the valley!

I remember the shambles during and long after the bombing raids, and as far as I can recall it was never cleared up in my time – no time, no men, no equipment, I guess.

I remember the black Dorniers, at tree-top height, dark and sinister. No wonder we needed no laxatives!

I remember a hit on a shelter and my first view of death. A girl's severed head, dirty and grey.

I remember digging and heaving that night by hurricane lamps, alongside a Salvation Army man, in vain efforts to aid the trapped bodies.

I remember the spreading claustrophobia as we sat in our shelters, and how we took instead to running through the woods behind dispersal, sprinting across the golf course, and diving into more woods to lie low until the all clear.

I remember the Salvation Army tea wagon and the magnificent service they gave.

I remember the famous red flags planted on the unexploded bombs – and our ignorance that they might go off!

I remember no spares or equipment, and the constant removal of props, CSUs, and boost controls and always adjusting the aneroid jets against continuous complaints of surge and low power.

I remember a raid at dispersal. I was changing a radiator with Sergeant Cockle when the sirens went. We worked on, and up roared a young pilot (Flying Officer Smythe?) insisting he had his machine, despite it not being ready – no cowl, no radiator fairing, and engine not run – the impossible handwinding of the Merlin as the bombers approached – no joy of course – so off rushed the pilot as we dived for our shelter. Last in, I found the bolt broken, and I held the door down. I can still feel the blast moving the door on my back, and its clang as I bounced it shut, again and again.

I remember one night the London skyline ablaze with fire, and an officer giving five rounds per man with orders to muster and repel the invasion!

I remember arriving at Acklington and the enormous relief at so normal and peaceful a scene.

Looking back after a lifetime in aviation I realise what a great big mess-up it all was. We won – but by luck as much as anything, plus the old British stamina!

Mr H Barker of Derby was a gunner on airfield defence before the formation of the RAF Regiment. He arrived in the spring of 1940 and remained until into the winter:

I spent the days before the Battle erecting and installing the gun posts around the airfield perimeter (the Army being in position outside), and I spent a lot ot time on the gun post erected on the roof of the Officers' Mess. We got to this post through a trapdoor in the roof, and I remember that after the barrack blocks were bombed we had to live, eat and sleep on top of a wooden-covered water tank in the loft of the mess.

I recall the long and sadly futile digging for survivors after a shelter got a direct hit, and I also recall a young NAAFI girl – Trixie – dashing to the canteen for 'some fags for the boys' at the beginning of a raid. She got the fags OK and started back when a Messerschmitt swooped down and machine-gunned her. He literally cut her to pieces before our eyes, and we all stood and cried.

The young pilots of 610 lived a day at a time – who could blame them – and they rip-roared about in sports cars, collecting road signs and street lights etc to display at squadron dispersal. One of the local pubs was called the White Bear and its sign was a full-sized Polar which stood in its garden. One night it went, naturally to appear in pride of place at 610s dispersal – but it was followed hot-foot by the landlord, and Group Captain Grice was not amused!

Mrs H M Stock of Watlington, Oxfordshire, was an administrative officer in the WAAF, and although not posted to Biggin Hill until just after the Battle (December 1940 or January 1941 – she cannot recall) she worked with Felicity Hanbury and remembers her well.

As Section Officer, Felicity Hanbury was Officer in Charge of WAAFs. It was her dedication, calmness and cool courage that set such a fine example to all the WAAF girls, and inspired them in turn to act likewise, and set a similar example to the men.

Felicity Hanbury was a young widow in 1940, her husband having been killed on active service late in 1939. Her courage, dedication and fine example during and after the Battle of Britain were recognised by the award of the MBE, the first such award to a WAAF officer during the war.

She remained in the service after the war, moving on to become Director of the WAAF from 1946-49, and Director of the reconstituted WRAF from 1949 until her retirement in 1950 as Air Commandant Dame Felicity Hanbury. Later she married Sir Harald Peake, the late Chairman of the RAF Benevolent Fund. Today, Dame Felicity lives in Oxfordshire and is still active in her work for the Benevolent Fund.

Like the three heroines on her 1940 staff, Dame Felicity's name is perpetuated at Biggin Hill as *Hanbury Drive*, in the married quarters site.

Mr B Durley of West Bromwich completed his basic

training at Melksham in Wiltshire and was posted to Biggin Hill as an AC2 in mid-August 1940. He recalls his dramatic arrival:

Enemy bombers were making a raid when we got to Bromley South Railway Station and we were dispersed in the transport that collected us. After the all clear we got to the Guardroom and we were checking in, thinking that all was safe, when the sirens went and we were ordered to take cover. I ran to a shelter to our right, and a lot of my pals dived for the one on the left. We were straffed with bullets by Jerry but the shelter to the left got a direct hit. What a mess when we got the all clear, and the awful task of helping to get the unfortunate ones out – this went on day and night.

The barrack blocks were bombed out so we had to make do in an old bombed Nissen hut, overrun with mice. Once settled, we were organised into No.75 Bomb Disposal Wing, but we knew nothing about bombs and learned the hard way by digging out the unexploded bombs and Ack-Ack shells. We would shout the fuse numbers back to our NCOs who would then come up and defuse them. After what seemed many weeks of bombing, I remember being billeted out at Keston, in Heathfield House, and I remember that Jerry followed us and bombed Keston – but not our billets. I soon got my AC1 and LAC – the pay as an LAC was useful – two bob a day, plus 3d for the bomb disposal badge!

Mr Durley's experience on arrival was shared by Mr C Carter of Bilston, Staffs, who must have been in the same party of new recruits arriving from basic training, for he writes:

I remember the welcome I received on arrival in mid-August with about 40 other lads. We had just checked in at the Guardroom when a raid began. I only had time to run to the side of the shelter before the bombs started. The planes were very low, you could see the expressions on the pilots' faces. One shelter received a direct hit.

The days that followed were spent filling up holes and clearing up, but there were many more raids. Then I was put on flare path duty. I went to the flare path billets and found a spare bed. An airman said it was the bed of one of the lads killed in the shelter – he had just come back from leave and had brought with him some white heather for luck.

There were more and more raids, and I remember that the best way to take cover was to lie low in the woods beyond the airfield. Later on, I remember we were billeted out at Heathfield House in Keston, and I was put on bomb disposal.

Mr W D Mumford of Hayes, Bromley, Kent, was a corporal in the station sick-quarters at the time of the August bombings. He still has a copy of the Medical Officer's report to the Station Commander, dated 4 September 1940, concerning the raids on 30 and 31 August. The report covers much administrative and medical detail and includes the following:

As the first bomb exploded the lighting failed; the emergency battery set was put on but that also failed after the third or fourth bomb. Emergency torches were procured and lack of lighting held up proceedings considerably. The dead were stored in a separate room and an orderly detailed to take particulars from their identity discs. The injured were seen on entrance by the MO, and an orderly was detailed to carry out the requisite treatment. In this way 22 casualties were treated in 45 minutes . . .

Word was then received that a shelter had been hit and help was required. An opening was found but it was plain that the majority of occupants had been killed. Those who were living, and within reach, were given a dose of morphia, several having their injections into the face as this was the only part of the man above the earth and concrete. Six living men were taken out but two died soon after. One of these latter had to have his wrist amputated in order to free him at all. The rescue squads worked hard and were relieved by ARP teams from Orpington, and then by a team of Welsh miners, upon whose arrival a great acceleration was noticed immediately. By 0200 hours most of the bodies had been removed . . .

A special commendation must go to Dr J C Colbeck of the Public Health Laboratory at Downe. He comes to our sick quarters for every raid and to date has missed none. When the MO was single-handed and 66 casualties were dealt with his services were invaluable. A tribute must go also to Doctors Grant and Mansi, both of Orpington, who worked for many hours performing amputations in cramped conditions and with enemy aircraft still overhead. Thanks also to the Orpington ARP, who sent two fully-equipped surgical units, and a dozen ambulances complete with nurses and drivers . . .

Dr Mansi now lives at nearby Green St Green, and still practices part time in Orpington, where his son runs a local practice. Today he praises the toughness and fortitude of the civilian population. "In those days, people did not bother doctors with trivial complaints," he says.

Mr R E Turrall of St Leonards-on-Sea, Sussex, was posted in during July 1940 and after a spell of duty on the Salt Box Hill machine gun post, was transferred to the Watch Office as airman of the watch. He recalls an incident which has been variously misreported:

On one occasion a Spitfire landed from Tangmere and taxied to the Watch Office for the pilot to report. To our great surprise we saw a WAAF climb down from the cockpit, followed by the pilot, Flight Lieutenant Bretell (92 Squadron). It was me who was sent to bring her to the Watch Office to face the music.

Bretell was subsequently court martialled on a charge of 'endangering one of His Majesty's aircraft', but after a spirited defence by fellow squadron pilots the charge was quashed.

Mr T H Meyler of Newport, Gwent, was Corporal in charge of the parachute section for 111 Squadron at Croydon, but as there was no parachute section at Croydon he operated from Biggin Hill. One of his frequent tasks was to take out a truck and tour around Kent picking up the parachutes after pilots had baled out. He writes:

If any of the pilots had been injured I would go along to the hospital to see them and then notify the Orderly Room of their condition. The first thing they would say to me was 'Don't forget my caterpillar Corporal!' I once picked a parachute up from the police station in Tunbridge Wells. The harness was practically burned away but it had managed to land its owner – and although he was very badly burned the first thing he said was 'Don't forget my caterpillar, and make sure it has red eyes!'

The caterpillar was the badge of the Caterpillar Club, membership of which was exclusive to aircrew whose lives had been saved by a parachute – the caterpillar badge being derived from the silk worm, parachutes being made from silk.

Mr A E Tidcombe of Clapham, London SW4, was a fitter with 92 Squadron. He has many proud memories among which are:

It was Wednesday, my day off. The Tannoy in the billet said 'All personnel to the shelters.' The chap in the next bed said 'B.....r the bombs, give me my tin hat!' He refused to go to the shelter. On returning from the shelter I found the ceiling down, and my comrade fast asleep in his bed, covered in plaster!

I remember the dawn patrols in the half light. Taking the wet covers from the aircraft, and the red glow of the exhaust pipes as the engines were warming up – all quiet and still, bar the deep roar of the revving engines. Wondering who would be coming back, as I had lost three aircraft on which I worked – and the grand lads who flew them.

On another occasion we were waiting for the return of the Squadron. A lot did not come back, including mine, but spotting a late straggler coming in and thinking perhaps it was mine, I waved it to dispersal. But it was taxiing too fast towards me and I dived in the grass to see it take off. As it passed overhead, I saw it was a Messerschmitt 109!

Mr H V Dean of Hove, Sussex, was an equipment assistant in the Petrol Dump from February 1940 until August 1941. He says:

I will never, ever, forget the petrol dump, its concrete control room and the stink of 100 octane. Many of my working colleagues were killed, including the little civilian who taught me, very thoroughly, all about the station petrol supplies. I remember the night after their deaths – sleeping alongside the wall of the shattered stores on the trusting assumption that lightning does not strike in the same place twice – and frightening a fellow airman who thought I was dead! . . .

I retrieved my buttonstick from the wreckage of the billet, and I have it in front of me now, bomb-scarred and hopelessly twisted. I am very proud of it . . .

I well remember the raid in which Ops Room was hit. I was returning from the dump when someone bawled at me to get across the road to see if help was needed. Me and another airman were the first to greet Group Captain Grice as he crawled from the wreckage, covered in dust and with blood running down his cheek. His first question was of the safety of the petrol dump. His next was an order to search the field, 'For I ordered some of my WAAFs to disperse there.' We found nothing except 'mole hills' which we viewed with deepening suspicion . . .

Memories flood back. I often wonder what those squadron commanders must have thought of the smelly, unwashed, tie-less young man who lived in the petrol dump and assured them that the petrol they signed for was as per bowser drivers' records and flying hours . . .

The 'scrambles' and take-offs, with aircraft in all directions like nests of eagles screaming upwards. The eerie noise of the fired guns as they came in to land. The feeling of utter helplessness as the ground swayed and writhed around the trench in a raid. Watching with crossed fingers as horribly damaged aircraft crept in and landed against all odds . . .

Bending the rule books and avoiding the forms in the stores system to get the supplies on time – and then the big sort out afterwards . . .

Geoffrey Howard of Warrington, Cheshire was an NCO fitter with 'A' Flight, 610 Squadron. He says:

I remember the day we arrived at Biggin from Prestwick. We were busy erecting storage tanks at our dispersal when the Station Commander appeared with a plump fellow dressed in a dark siren suit. Yes, it was Churchill, and the lads heard him say, 'So this is the Cheshire Squadron!'

I recall vividly the Sunday Raid. 610 and 32 had 'scrambled' and the Tannoys told us all to take immediate cover. At the aircraft dispersal bay we had a sandbagged enclosure with slabs of concrete on top, and we could move along inside this on our knees. I found that the birds gave the first sign of imminent danger, complete stillness – not a sound, and no movement of birds in flight – uncanny. Suddenly a smoke bomb slap-bang in the middle of the airfield, then sticks of bombs came hurtling down. And afterwards all was quiet again, the casualties taken care of and the fires burning out. A single jeep was moving on the airfield, two people in it, one a WAAF. The jeep was loaded with steel spikes carrying red flags, and as it passed the mounds of earth, the WAAF would jab a spike into the mound. They carried on until the whole area was like a poppy field. They were marking unexploded bombs!

Norman Gilham of Capel, Dorking, Surrey, was a cadet in the local Air Training Corps. In a letter he recalls:

We used to go to Biggin Hill at weekends and help with various tasks such as loading the ammunition belts into guns, and polishing the aircraft – for we were told that a really clean 'kite' could fly up to 10 mph faster.

I well remember the thrill of seeing the 'kites' coming back. The first thing we looked for was the tattered 'elastoplast' over the gun ports – telling us that the pilot had been in a scrap, and this was the case more often than not.

The sight of odd and assorted chairs littering the grass in front of the squadron offices, without occupants and with no aircraft in sight, stands out in my mind. It must have been the tenseness in waiting for the occupants of those chairs to return that made the memory so indelible.

Pilot Officer I H Cosby joined the Service in 1938. His first operational role was as an Army co-operation pilot with the BEF in France. On return to England in 1940 he volunteered to be a fighter pilot and was given six days to learn to fly the Spitfire before joining 610 Squadron at Biggin Hill in September 1940. Later he joined 72 Squadron and then became a night-fighter pilot on Defiants with 141 Squadron. After the war, having won the DFC, he completed numerous staff and flying appointments before returning to Biggin Hill in 1967 as a Wing Commander on the staff of the OASC, and then completing his service with a tour on the Air Staff in Whitehall. Now retired, he lives in Biggin Hill village and retains close ties with the Station as an honorary member of the Officers' Mess and an active supporter of St George's Chapel. Today he remembers something of the flying conditions of forty years ago, and he vividly recalls two combat reports:

Our attire in 1940 was somewhat irregular to say the least. Although we were all issued with fleece-lined leather jackets, known as 'Goon Skins' after the Popeye cartoon of the time, they were far too bulky to wear under our Mae Wests which were filled with kapok. This left us with basic uniform tunic and slacks, plus fleece-lined flying boots. This was OK at ground level when sitting around at dispersal, but when launched into space at very short notice, it was only a matter of minutes before we were hurtling through the sky at twenty-five to thirty thousand feet with the temperature the same as at the top of Mount Everest. Moreover, the canopy had a habit of misting and icing up and we frequently had to slide it open, and you can imagine how a 300+mph wind rushing past added to our discomfort. I resorted to laddered silk stockings and silk pyjamas under my uniform and a woolly roll-neck 'Flash Harry'. A compromise arrangement to meet the extremes of temperature, but even then it was terribly cold up top, one's hands became quite numb – despite wearing leather gauntlets over silk inner gloves.

Two scraps come to mind! The squadron was jumped by a flock of Me109s over the South Coast. They came out of the sun and the first I was aware of them was a stream of tracer passing over my head and ripping pieces from the tail of the Spitfire fifteen feet in front of me. The whole formation broke station and we mixed it in a gigantic ball of aircraft, turning, weaving, climbing and diving, with tracer bullets flying everywhere. Suddenly a Me109 swerved round in front of me, so I let him have a short burst from my eight machine guns. Immediately, one of his undercarriage legs dropped down. Another burst of fire, and white vapour streamed back from his engine – I had got his coolant system. He went into a steep dive and I followed him down, keeping a wary eye on my mirror to ensure I didn't get the same treatment. He levelled out along the coast and lost height over Folkestone harbour. I gave him another burst and he went into the sea with a great splash. I wheeled round and did a victory roll over the patch of white foam before heading back to Biggin. On landing I discovered a large hole in my engine cowling, a 20mm cannon hole in my *coolant tank, but enough had remained in to get me back. Needless to say, I never repeated the victory roll bit– I could have lost* all *my coolant! I'm told the Me pilot survived because an Army officer dived from the Mole, swam to him and brought him ashore.*

With 72 Squadron over the Thames Estuary I was acting as the 'weaver' – ie. weaving back and forth above and behind the squadron, keeping a look out to the rear – impolitely known as 'Arse-end Charlie'. We had been vectored to a large formation of bombers heading for London, and we saw them crossing ahead and below, but as they were under AA fire we waited for them to break formation before we attacked. One-by-one our sections moved in and I was left on my own. Looking up I saw a long line of 109s crossing me from port to starboard and coming out of the sun. The rule was 'Never start a fight on your own – get down and get home.' But there is Walter Mitty in all of us – there certainly was in me that day – and I counted twenty-eight in this loose line. I visualised climbing up behind and working my way along, picking them off one at a time. Swinging into a steep climbing turn, I got my gunsight on to what I thought was the last one. Before I could open fire my canopy shattered with a loud bang and I was upside-down and dangling in my straps. The rush of air threw pieces of grass and caked mud from the cockpit floor into my face and eyes, and my aircraft went into a spin. The Estuary and the South Coast spun round in front of my cowling, I was strangely detached from it all – like an onlooker. Pulling myself together, I took the controls and forced the aircraft out of its spin and into a dive. Looking round I saw that my port aileron was shot away, there were large holes all along the wing, and my mirror was gone. Clearly Me109 number twenty-nine had been waiting for just such as me and I had obliged! Fortunately he did not follow me down and I limped home to Biggin.

There were many scraps at Biggin, and later at Hornchurch and Tangmere but just after Christmas I moved on to Defiants at Gravesend. But that is another story!

And so their tales are told – or some of them. For sixteen weeks in 1940 Britain was in desperate crisis. She survived because the Royal Air Force refused to lie down and be beaten. Embodied in that refusal was the spirit and the will of the people: the pilots, the fitters, mechanics, controllers, drivers, administrators, caterers, the WAAFs and the civilian personnel, and the brave local people bombed out of house and home by the 'ones that missed'.

It was not Biggin Hill alone, but the whole Sector, and the other sectors, 11 Group, 12 Group, and 10 and 13, Fighter Command, the Army, the Navy – the Nation. The Biggin Hill story is rightly renowned, not just in itself but as a focus; for it is also the Kenley Story, the Tangmere Story, Hornchurch, Manston, North Weald . . . They all had their *people*, and together we won.

Today we remember. We remember with pride and thanksgiving. We remember with sadness, and with humour and joy.

The Squadrons

Seven squadrons fought in the Battle from Biggin Hill. Most had their origins in World War I, and after 1940 they moved on from Biggin to distinguish themselves in almost every theatre of the Second World War. Three of them still fly today. The outline histories of the seven squadrons are told on the following pages.

Hurricanes of 32 Squadron, Empire Air Day, 1939.
(Photo: RAF Biggin Hill)

32 Squadron

32 Squadron was formed from surplus elements of 21 Squadron as a Royal Flying Corps unit at Netheravon in January 1916. It first saw action as a scout squadron flying the DH2 pusher fighter against the Fokker monoplane over the trenches late in May 1916. On 1 July, the opening day of the great Somme offensive, the squadron won its first battle honours when the CO, Major L W B Rees, was awarded the VC after making a single-handed attack on a formation of ten German fighters, breaking their formation and forcing two of them to the ground.

The squadron stayed in France for the remainder of the war, patrolling over the trenches during the dreadful stalemate of 1916/17, and re-equipping with the DH5 in the early summer of 1917. It flew the DH5 in the ground attack role at the Third Battle of Ypres, and it continued in this role until the SE5a replaced the squadron's DH5s in January 1918.

32 took its SE5s into action against the big German offensive of 1918 and then briefly transferred to the French front, returning to British lines for the Battle of Amiens and the German retreat.

After the Armistice the squadron returned to England, to Tangmere, in March 1919, and then moved on to Croydon where it disbanded in December.

All Fools' Day 1923 saw the squadron's rebirth at Kenley, equipped with Sopwith Snipes. The squadron remained at Kenley until September 1932 when it moved across the valley and the county border into Kent and settled at Biggin Hill. During its near decade at Kenley it was equipped successively with Snipes, Grebe IIs, Gamecock Is, Siskin IIIas, and Bulldog IIas, and at Biggin it flew Gauntlet IIs before taking delivery of some of the early Hurricanes in October 1938.

By the outbreak of World War II the squadron was fully worked-up on the Hurricane I and was fit and ready to face the fury of the Luftwaffe. The 'Phoney War' produced many 'scrambles', mainly abortive, and found 32 operating out of Gravesend and Manston as well as Biggin, but by early June the squadron was fully involved in the battle over France. From mid-June onwards the pitch of fighting steadily increased as the Luftwaffe launched its pre-emptive raids on the airfields of Kent and South-East England.

32 bore the full brunt of the attacks; operating under the leadership of squadron leaders John Worrall (later Air Vice-Marshal) and Michael Crossley, it was constantly in the air during the long daylight hours of July and August, its busiest spell coming with the heavy German bombing raids in mid-August. At the end of this month 32, having been at the forefront of the heaviest and most prolonged engagement in the then history of air warfare, was sent north to Acklington for a well-earned rest, with one hundred and two German aircraft kills to its credit.

The squadron then started night-fighter operations before moving south again, to Middle Wallop, in December. There followed a relatively inactive period in the west country on convoy patrol but by November 1941, 32 was back in Kent operating out of Manston on fighter sweeps and escorts. The squadron then moved on to West Malling in May 1942. Whilst at Malling, 32 took part in the Dieppe raid of 19 August, providing low-level air cover over the beaches throughout the hours of daylight, and meeting for the first time a troublesome adversary in the new FW190. In September the squadron was withdrawn from operations to train for the North African landings and in December 1942, 32 Squadron left the UK for Phillipeville to start a period of some 26 years continuous overseas service. Conversion to Spitfires started in April 1943 and 32 saw out the war in very active roles on the various Mediterranean fronts, including Naples, the Adriatic and Dalmatian Coasts, Jugoslavia and Greece, and the Salonika landings.

32 then moved to Palestine until 1947 where it policed the skies on internal security duties, following this with the same tasks in Cyprus, still equipped with Spitfires, before moving on to the Canal Zone and converting to its first jets, Vampires, in 1951.

The squadron remained in the Mediterranean as a fighter-ground-attack unit of the Middle East Air Force and re-equipped with Venoms, the Vampire's more potent successor, in September 1954. Venoms were to be 32's last fighters and it operated with them from Shaibah, Amman and Mafraq until January 1957 when the squadron was reconstituted as a light-bomber unit equipped with Canberra B2s and operated out of Nicosia before moving to its more permanent base at Akrotiri.

32 spent the next 12 years in Cyprus before disbanding in 1969 and then returning to UK to reform at Northolt as the new Strike Command Communications Squadron. It remains in this role today, still at Northolt and operating a wide variety of communications (light and medium passenger) aircraft, including the sleek DH125 'executive' jet, Andovers and Gazelle and Whirlwind helicopters. In this role it is responsible for transporting Government Ministers, senior officers and other VIPs around the UK and throughout Western Europe, and occasionally further afield.

The squadron badge is a stringed hunting horn, depicting the traditional role of hunting the enemy. The badge was approved by King George VI in December 1936, and the motto – 'Adeste Comites' – is translated as *Rally round, comrades.*

Gladiator of 72 Squadron at Farnborough, 1937.

72 Squadron

Very few squadrons are fortunate enough to lead continuous and uninterrupted lives – at some time or other most suffer from the vagaries of operational requirements and are caused to disband or re-role. 72 (Basutoland) Squadron is no exception and has been disbanded and reformed on no less than three occasions.

The squadron was a late addition to the Royal Flying Corps, and was formed at Upavon in July 1917 from a flight of the Central Flying School. Early in 1918 it moved to Mesopotamia and its three flights dispersed, each operating different aircraft: 'A' Flight with SE5as, Spads and DH4s at Samarra, 'B' Flight with G-100 Martynsydes (Elephants) at Baghdad, and 'C' Flight at Mirjana with the Bristol monoplane. 'C' Flight used the monoplane to good effect in the capture of Kirkuk, and in one celebrated incident overawed a whole Kurdish tribe by performing low-level aerobatics immediately above their heads and causing them to cross to the British side – such was the influence of early air power!

Meanwhile, 'B' Flight's Martynsydes were deployed with the Army 'Dunsterforce' in the break for the Caspian Sea in order to prevent the Turks from entering Persia, and 'A' Flight was operating against the Turks on the Tigris front.

On completion of these operations the three flights rejoined at Baghdad, and in September 1919 the squadron disbanded after a short but most interesting war.

72 did not reform until February 1937 when it split from a flight of No.1 Squadron and took delivery of the first production Gloster Gladiators at Tangmere. Its CO on reforming was Flight Lieutenant E M Donaldson who was later to hold the World Air Speed Record in 1946, and who retired as an Air Commodore to become the Air Correspondent of *The Daily Telegraph*.

The squadron then moved north to Church Fenton as the first of many operational Gladiator squadrons. The Gladiator, however, was something of a stop-gap before the quantity deliveries of the high-performance monoplanes and, in April 1939, 72 received its Spitfires – just in time to reach full operational readiness with these machines by the outbreak of war.

The first few months of the war were spent in the north-east, and in Scotland (at Leconfield, Drem and Acklington), mainly on convoy and coastal patrols, but in

June 1940, 72 spent a brief few days at Gravesend helping to cover the desperate evacuation from Dunkirk. The squadron then moved back to Acklington before coming south to Biggin late in August at the height of the Battle of Britain. By the end of September its claims amounted to fifty-three and a half enemy aircraft destroyed, fourteen probably destroyed, and twenty-six damaged – not a bad month's work!

72 spent a quiet winter back in the north-east, and in April 1941 it re-equipped with Spitfire IIs then, in July Spitfire Vs, before moving south again to Kent – this time for a spell of one year operating alternately out of Gravesend, Biggin and Lympne. This was a hectic year with intensive operations on fighter sweeps and bomber escorts, including escorting the Swordfish which torpedoed the *Scharnhorst* and the *Gneisenau* in February 1942.

In the late summer of 1942 the squadron flew north again to train for overseas, and then left the UK for Gibraltar in November, quickly moving on to Algeria in support of the North African campaigns, where it served with distinction, scoring a further fifty-three kills.

With the successful conclusion of the Allied operations in North Africa, 72 moved on to Malta in June 1943 for the invasion of Sicily, and then onwards into the Italian campaign. Its two-hundreth kill was achieved in September, and in 1944 the squadron gave fighter cover for the Anzio landing and for the invasion of southern France. 72 then re-equipped with Spitfire LFIXs to become a fighter-bomber unit operating in support of the 8th Army. Its work in this role included a brief spell in Austria in 1945, and its final wartime base was Zeltweg from where it flew border patrols with the occupation forces and undertook various policing activities before disbanding at Zeltweg on 30 December 1946.

The squadron then reformed with Vampires at Odiham on 31 January 1947, when 130 Squadron was renumbered 72. In March 1950 the squadron moved to North Weald with Vampire FB5s and then re-equipped with Meteor F8s in 1952, before returning to its old northern stamping grounds at Church Fenton, where it flew Meteor NF12s and 14s. This highly mobile unit moved on yet again in September 1959, this time to Leconfield where it operated with Javelins until June 1961, when it once more disbanded and finally closed the pages of its history as a fighter squadron.

But 72 could not be kept down, and in 1962 Odiham was once more the scene of its rebirth – this time as a helicopter unit in 38 Group. The squadron remains there today, flying the Wessex, as a very busy and distinguished close support unit – still within 38 Group.

The squadron badge, approved by King George VI in 1938, is a swift volant, depicting speed – and the motto needs no translation. The name 'Basutoland' derives from the pre-war Commonwealth Air Training Scheme.

74 Squadron

74 Squadron's Lightning F1s lined up at Coltishall in 1961.
(Photo: Ministry of Defence)

Like 72 Squadron, 74 (Trinidad) Squadron was a late addition to the Royal Flying Corps. It was formed as a training unit at Northolt on 1 July 1917 and its first machines were SE5s. The squadron operated in the training role from London Colney before moving to France in March 1918, arriving just after the German offensive.

The squadron destroyed five German aircraft without loss in its first combat engagement on 12 April 1918, and for the final months of the long war it operated on offensive patrols dropping 25lb bombs on opportunity targets. At the same time it played a very effective part as a fighter squadron and by the Armistice it had destroyed a total of one hundred and forty enemy aircraft. 74 then returned to the UK in February 1919 and was disbanded at Lopcombe Corner in July.

The squadron was reborn in September 1935 and the event was unusual in that it occurred aboard a troopship – the *Neuralia* – en route to Malta. 23 Squadron, then at Biggin Hill, had a detachment in Malta at this time as a precaution during the Abyssinian crisis and as change-over personnel were embarked it was decided to expand this detachment into a fully independent squadron; so 74 was flying again, this time with Demons out of Hal Far. But in August 1936, after only a year in Malta, 74 was homeward bound to Hornchurch to become part of the London Defence Force, and later to re-equip with Gauntlets, and then, early in 1939, with Spitfire Is.

At the outbreak of war the squadron was still at Hornchurch and was ready for action, although for a while it was dogged by ill-luck – especially in September when it suffered the dreadful misfortune of shooting down two Hurricanes of 56 Squadron, momentarily mistaking them for Germans. The winter 1939/40 – 'The Phoney War' – brought little action and it was not until May that 74 was flying in anger once more, this time over France where, in the busy and desperate confusion of the retreating armies, it lost five pilots in the space of a few days.

Back at Hornchurch in June, it soon overcame its earlier misfortunes and from July it was heavily involved in the Battle of Britain, moving north to 12 Group in August. A brief spell at Coltishall preceded its move to Biggin, where it arrived in

October under the command of the legendary Squadron Leader 'Sailor' Malan. It came with Spitfire IIs to replace 72 Squadron who had moved north to rest at Leconfield.

Although 74 was a later arrival as far as Biggin was concerned, it went wholeheartedly into the fray and by the end of 1940 the squadron had run up a formidable score.

1941 brought a variety of locations ranging from South-East to North-East England, North Wales and Northern Ireland, and an equally wide variety of operations including convoy patrols, bomber escorts, and fighter sweeps.

In April 1942 the squadron moved out to Palestine but on arrival it received no aircraft and was held in reserve for some months before moving on to Iran with Hurricane IIbs and then, in August 1943, to Egypt to become operational with Spitfire Vs. Many moves around the eastern Mediterranean followed with relatively little action and in April 1944 the squadron came back to England to re-equip with Spitfire IXs at North Weald.

74 then moved back into Kent and from Lympne it was very active over the D-Day invasion beaches before moving on to France as a Spitfire-bomber squadron within the newly formed Second Tactical Air Force. It operated in the close-support ground-attack role, hedge-hopping along through France and Belgium with the advancing Allied armies.

At the end of the war 74 moved to Colerne to join 504 and 616 Squadrons to form the RAF's first jet fighter wing, flying with the new Meteor F3.

The squadron's long-term post-war home then became Horsham St Faith in Norfolk from where it flew various marks of Meteor until replacing them with the Hunter in March 1957, and moving to Coltishall in June 1959. After a year at Coltishall it achieved another notable 'first' in receiving Lightning FIs and becoming the first Mach 2 fighter squadron in RAF service. As the Lightning force expanded, 74 moved north to Leuchars for a three-year spell in Scotland.

June 1967 brought yet another 'first' when the squadron flew its Lightnings out to Singapore to operate out of Tengah and become the first supersonic squadron in the Far East Air Force.

Alas, the British forces were withdrawn from East of Suez early in the seventies and 74 was disbanded at Tengah on 1 September 1971.

So the Tigers fly no more, but their distinctive badge – the snarling tiger's head with the motto *I fear no man* – remains as a lasting reminder of their many sacrifices and achievements. 74 was also in the Commonwealth Air Training Scheme as 'Trinidad' Squadron.

79 Squadron

Three men link up to swing the prop of a 79 Squadron Dolphin in 1919.
(Photo: Imperial War Museum)

79, 'The Madras Presidency' Squadron, was another unit formed in the later months of the Royal Flying Corps, and was born at Gosport in August 1917, as a training squadron. It soon took delivery of Sopwith Dolphins and moved to France in February 1918 where, although hardly ready for battle, it was thrown in to the melee of the great German offensive. The squadron moved rapidly from base to base (or rather from field to field) and in the space of a few hectic months underwent a baptism of true fire, coming through with a total of sixty-four enemy kills, but these achieved at the expense of heavy squadron casualties. After the Armistice, 79 joined the Army of occupation in Germany before being disbanded at Bickendorf in July 1919.

79 was reborn out of 'B' Flight of 32 Squadron at Biggin Hill in March 1937 and was then equipped with Gauntlets. Hurricane Is were received in November 1938 and with these it entered the war, seeing its first action over the Channel whilst operating out of Manston in November 1939. The war's first winter was quiet and dull, but in May 1940 the squadron went to France for an intensive ten-day action covering the Army's desperate retreats, and in this brief but hectic period 79 moved base three times and claimed twenty-five enemy aircraft destroyed.

Back in England, the squadron fought on as an 11 Group unit before moving up to Acklington in July for a brief spell as part of 12 Group's northern defensive force. Late August saw 79 back south at Biggin as a replacement for 32, arriving at the height of the repeated bombing raids on the station. The squadron was under constant pressure and was operating out of the chaos of the bombed-out runways and hangars, and on one occasion early in September, when airborne during a raid on its beloved base, it had to divert to Croydon on return, being unable to land on the shattered runways at Biggin. Despite its short stay, 79 acquitted itself with customary distinction before being moved on to rest at Pembrey, South Wales, in mid-September.

The squadron remained in Wales for many months and worked on convoy escorts before re-equipping with Hurricane IIas and IIbs at Fairwood Common late in 1941.

With the entry of Japan into the war there arose an urgent need for fighter squadrons in the Far East, and in December 1941 79 moved to Baginton to receive Hurricane IIcs and embark for India. The squadron arrived in Bombay in June 1942, but did not see action until December when its detachment at Chittagong accounted for five Japanese bombers from a group which raided the airfield.

1943 was spent in ground-attack and bomber escort missions across the eastern front, working mainly from Comilla in the Ganges Delta, and Ramu close to the Burmese border. Later in the year the squadron moved back into India for ground-support training, and was not in action again until the winter when it flew many escort missions to Dakotas on supply drops to the besieged British troops in Burma.

1944 saw the squadron back supporting the Army in the ground-attack role, and in May, after re-equipping and training with the American Thunderbolt, 79 returned to the war as a tactical fighter-bomber unit to make a major contribution to the British offensives in the spring of 1945 which cleared the Japanese from Burma. The squadron then disbanded at Meiktila at the end of December 1945.

79 remained disbanded until November 1951, when it reformed at Buckeberg in Germany as a fighter-reconnaissance unit within the Second Tactical Air Force, equipped with Meteor FR9s and later, very briefly, with the short-lived Swift FR5s. The squadron remained in Germany in this role for ten years before finally disbanding in 1961.

The squadron badge was approved by King George VI in March 1938. It depicts a salamander salient in its flames. The salamander, of course, is fearless, and the motto 'Nil nobis obstare potest' is translated as *Nothing can withstand us* – a justifiably defiant boast.

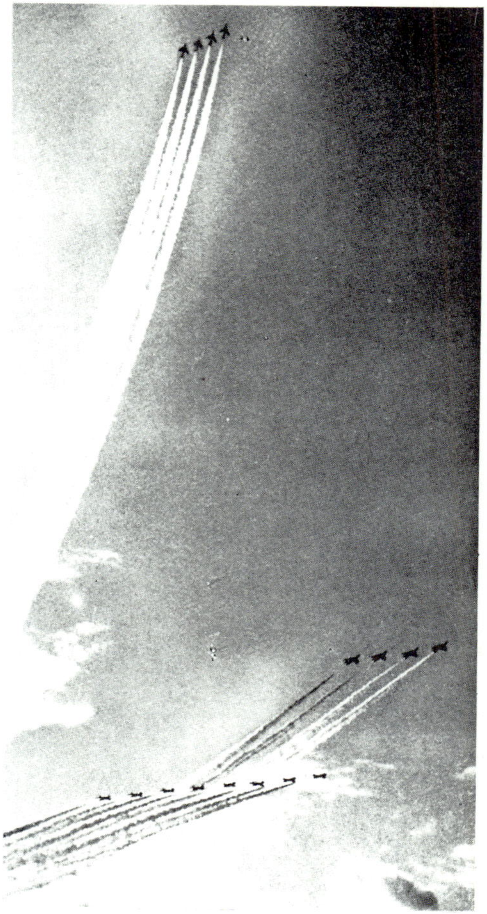

92 was the official aerobatic team in 1962 – *The Blue Diamonds* – equipped with 16 Hunter F6s.

92 Squadron

Like so many of the early squadrons, 92 also originated as a Royal Flying Corps unit. It was formed at London Colney in September 1917 as a scout squadron equipped with Pups, Spads and SE5as.

In July 1918 it moved to France with its SE5s and operated in the Dunkirk area and from many bases along the Western Front. The squadron took part in the Somme offensive and was very busily employed until the end of the war, by which time it had a score of thirty-seven enemy aircraft destroyed – a formidable total in so short a time and certainly an early portent of even greater things to come, but with the war ended, 92 disbanded at Eil in August 1919.

The squadron remained 'on the shelf' until October 1939 when it reformed at Tangmere out of Auxiliary and Volunteer Reserve personnel of 601 Squadron. It was briefly equipped with Blenheim Ifs before receiving Spitfire Is at Croydon in March 1940. 92 was quickly in action over France and the English Channel, and in late May it destroyed six 109s on its first operational patrol over the French coast. The squadron quickly gained a reputation for cool efficiency, a casual flambouyant approach which belied dedicated professionalism and a firm (if not apparent) discipline, and in June, after prolonged and intensive French and Channel patrols, 92 withdrew to South Wales for a change of scene. On 9 September, 92 came to Biggin to join the Battle of Britain as replacement for 79 which had been pulled out to rest.

The squadron – 'The 92nd Foot and Mouth' – was soon to become perhaps the best known unit in Fighter Command and was thrown immediately into the thick of things at the *Bump*. By the end of 1940 its score was one hundred and twenty-seven enemy aircraft destroyed – by quite a long way the highest bag at that stage of the war.

1941 brought re-equipment with Spitfire Vbs and in the autumn the squadron left the UK for Egypt, where its arrival without aircraft (not its fault!) caused the pilots to operate with 80 Squadron's Hurricanes until August 1942 when 92's Spitfires arrived. The squadron then gave valiant and successful offensive support against Rommel's advance on the Nile when it beat back repeatedly large numbers of German and Italian fighters. It then supported the British stand at El Alamein which led to the successful Allied counter-offensive and the speedy rout of the German Africa Korps. 92 was involved in the hasty rush from desert strip to strip, and its operations included Tripoli, the Mareth Line and the advances into Tunisia.

By June 1943, 92 had moved to Malta to cover the Sicily invasion and the landings at Pachino. It moved on into the Italian campaign, patrolling over Messina, and by 14 September it was flying from Italian mainland soil. 92 then moved up the Adriatic coast with the Army advance and in early 1944 it crossed to the west coast to cover the Anzio landings. For the remainder of the Italian campaign, the attack on the Gothic Line, and the landings in southern France, the squadron operated in the fighter-bomber and ground-attack role before moving on into Europe and into Austria as part of the occupation forces.

It disbanded at Zeltweg on 30 December 1946, and it finished its war as the RAF's highest scoring squadron with three hundred and seventeen enemy aircraft to its credit.

Disbandment was brief for after only a month, on 31 January 1947, 92 was back in business at Acklington, having reformed out of 91. The squadron's new aircraft were Meteor F3s and it flew these from Duxford in the air defence role until May 1948 when it re-equipped with F4s, before moving on to Linton-on-Ouse. The next eight years were spent at Linton, operating successively with Meteor F8s, Sabre F4s and Hunter F4s, still as an air defence squadron. Middleton St George then became the squadron's home from March 1957 until May 1961, with a break at Thornaby in 1957/58, and from Middleton it flew the Hunter F6.

92 then moved on to Leconfield, and early in 1963 it became supersonic on converting to the Lightning F2. December 1965 brought a move to Gutersloh in RAF Germany, and today 92 remains in Germany operating out of Wildenrath, flying Phantoms, still in air defence – the pure fighter role.

The squadron badge is a cobra entwining a sprig of maple leaf. The maple signifies the squadron's early Canadian associations, and the cobra is in recognition of the Commonwealth Training Scheme in which 92 was one of the 'East India' units. The motto 'Aut pugna aut morere' – *Either fight or die*, is perhaps the inspiration of 92's successes – for it has always fought hard and it is very much alive.

141 Squadron

The formation of 141 Squadron was authorised in the autumn of 1917; the squadron was to operate from Biggin Hill as a unit in the London Defence Force and was to come in to being on 1 January 1918. Meanwhile, London was suffering from the sporadic raids of German Gotha bombers and the south of the capital could not be left undefended. A flight of 39 Squadron at Hornchurch was therefore dispatched to Biggin to hold the fort until the arrival of 141. Thus 141 Squadron became the first full squadron to operate from Biggin Hill.

141's early aircraft included a mix of Sopwith Dolphins and Pups, BE12s, Vickers FB26s and BE2s, before it settled for the Bristol F2b in March 1918. In May the squadron claimed its first, and only, victim of the First World War when it downed a Gotha over the North Downs east of Maidstone. The squadron saw no further action until the end of the war when it became one of the few squadrons ever to serve in Southern Ireland where, during the latter stages of the unrest, it operated from Tallaght, Curragh and Baldonnel between March 1919 and February 1920. The squadron was then disbanded at Baldonnel on 1 February 1920.

It did not reappear until October 1939 when it was reformed at Turnhouse. At first, it had no allocated aircraft and it worked-up on any machines available: Ansons, Blenheims, Tiger Moths and Gladiators etc, before receiving Defiants, the new two-crew fighter with pilot and a gunner who operated from a mid-fuselage turret, and becoming operational in June 1940.

At this time 141 was one of only two Defiant squadrons. The other, 264, had taken the aircraft to Dunkirk where it met with only limited success and proved vulnerable to enemy attack. The lack of forward-facing fixed machine guns was shown to be a serious shortcoming in the use of this aircraft as a day fighter, for the Germans soon realised that if attacked from beneath and behind the gunnere could not see approaching raiders and the aircraft was a sitting duck. Consequently, Defiants were withdrawn and used only as night-fighters. Nevertheless, the desperate shortage of aircraft during the Battle of Britain caused 141 to briefly return to the interceptor role when for some two weeks in July 1940 they were back at Biggin Hill on Channel patrol. They were all but wiped out, however, and this repeated failure dictated their permanent allocation to night fighter duties.

The squadron met with success in this role and later, in September, towards the end of the Battle of Britain, it was back at Biggin Hill for a few days during which it shot down two Heinkel 111s and a Junkers 88 on night patrols. 141 then returned north to continue successful night operations, flying from Ayr and Drem in Scotland and from Acklington. Conversion to Beaufighters came in August 1941 but patrols were uneventful and it was not until the squadron moved south to Tangmere in June 1942 that more successes were scored. 141 then changed role to make daylight intruder raids on trains and ships and to act as escort to Coastal Command shipping raids. In mid-1943 it received the Beaufighter VIf; with this longer range machine it was able to extend raids over the Dutch coast, and during this period operated it from Wittering. December 1943 brought conversion to Mosquito IIs and a move to West Raynham, from where it flew as a 100 Group, Bomber Command unit on bomber support duties, but it also continued intruder raids, particularly over Holland and Belgium, with the Mosquito. 141 finished the war back in the night-fighter role with Mosquito NF30s and was disbanded at Little Snoring in September 1945.

June 1946 saw its rebirth at Coltishall, equipped with Mosquito NF36s and still in its familiar role. The squadron remained at Coltishall for the rest of its career, converting successively to Meteor NF11s, Venom NF3s, and lastly, the Javelin FAW4. Disbandment came again on 16 January 1958 when 141 was 'taken over' by 41 who had that day disbanded at Biggin Hill and moved to Coltishall to fly 141's Javelins in their own name. In this way, 141 Squadron was the first and very nearly the last squadron to fly from Biggin Hill.

141's badge is a leopard's face on an ogress, symbolising fighting at night, and the motto 'Caedimus Noctu' is translated *We stay by Night* – a motto brought vividly to life by a distinguished squadron which spent some fifteen years on night-fighter operations.

A 141 Squadron Javelin FAW4 in 1957.
(Photo: Ministry of Defence)

610 Squadron

610 (County of Chester) Squadron was an Auxiliary unit and started life in February 1936 at Hooton Park in Cheshire as a light bomber squadron equipped with Hawker Harts, Avro Tutors, and later with Hawker Hinds. On 1 January 1939, 610 was transferred to Fighter Command, but the sudden rapid expansion of the Air Force and the consequent shortage of aircraft caused delay in the arrival of the squadron's first fighters, and some confusion as to which type it was to receive.

At first, it was designated as a Defiant unit, but by September no Defiants had arrived and so for a period of three weeks the squadron flew Hurricanes before taking delivery of Spitfire Is during October. 610 spent the winter of 1939/40 as a 12 Group unit based at Wittering, and worked-up to operational perfection.

The spring of 1940 brought a brief spell at Prestwick, then early in May the squadron was moved south to Biggin Hill as replacement for 79 Squadron in the actions covering the British Army's retreat from France and evacuation from Dunkirk. This was 610's first taste of real action and the squadron was remarkably successful, for in the three weeks or so from mid-May until early June, 610's Spitfires were on regular patrol, often two or three times daily, over the Channel and the Dunkirk beaches. Their score in this period was eight kills and four probables – but with the loss of six of their own aircraft – half of the squadron, and on 4 June they were withdrawn to Gravesend to rest and re-equip.

Spitfire Is of 610 Squadron patrol over Kent in June 1940.
(Photo: Imperial War Museum)

Later in July, 610 returned to Biggin Hill where, together with 32 Squadron, it bore the brunt of the Luftwaffe's increasingly heavy raids on the RAF in the south-east – the build-up for Hitler's 'Operation Sea-lion', and prolonged action that was to become known as the Battle of Britain. Throughout August, 610 was almost constantly in the air – sometimes 16 hours in a day – against seemingly overwhelming odds, and perhaps its finest hour came on the 18th – the hardest day – when it intercepted a massive formation of German bombers and prevented what would probably have been the heaviest raid ever on Biggin Hill.

The tremendous pace of the battle began to tell on men and machines, and at the end of August, 610 was pulled out to rest and recuperate at Acklington in the 'quiet north'. 610 then spent a few months on North Sea convoy duty before coming south again in December 1940 to join the Tangmere Wing at Westhampnett. Here it received Spitfire IIs and was engaged mainly in offensive sweeps over the Channel, before moving up to Leconfield to rejoin 12 Group on shipping patrols once more. At Leconfield it received the cannon-firing Spitfire Vb and with these it concentrated on bomber escort work, but 610 also took part in the Dieppe raids in August 1942, operating temporarily out of West Malling for this task.

1943 brought a variety of operations and locations, including a spell in Cornwall and many fighter sweeps and raids over France. January 1944 found the squadron based at Exeter and here it became the first unit to receive the Griffon-engined Mk XIV Spitfire – the fastest version to date. After working-up with this new machine the squadron moved eastwards to West Malling and Ludham where the Mk XIV's speed was put to good use in a three-month spell of combat against the pilotless V1 'Doodle Bugs' which were menacing our towns and cities from their many launching sites along the French coast. By late August 1944, when the V1 threat was over 610 had bagged a total of fifty of these fearful menaces.

The squadron then put drop-tanks on its aircraft and undertook long-range sweeps over Germany, rooting-out the Luftwaffe from wherever it could be found, and in December, 610 joined the Second Tactical Air Force at Evere in Belgium. With 2 TAF it continued with long-range sweeps and moved on with the advance towards Germany, returning to the UK for a break in February 1945. Allied progress was so rapid at this stage that 610's return to the Continent was not required and the squadron was then disbanded at Warmwell in March 1945.

July 1946 brought its reformation as an Auxiliary squadron, back at Hooton Park, its original home. Its first post-war machines were Spitfire XIs again, but it soon received the Mk F22 before 'modernising' on to Meteor F4s and then F8s. 610 remained at Hooton until it disbanded early in 1957 when the Auxiliary Air Force lost all its squadrons and was reduced to a small maritime headquarters support formation.

The squadron badge is a garb or wheatsheaf and is taken from the armorial bearings of the City of Chester. The motto 'Alifero tollitur axe ceres' is translated as *Ceres rising in a winged car.*

St George's Chapel of Remembrance

Whatever the future holds for the RAF Station, St George's Chapel of Remembrance will remain as a permanent memorial to the aircrew of the Biggin Hill Sector who lost their lives in the Second World War.

In the Shadow of Thy Wings Will I Rejoice . . .

Psalm 63

St George's is not a Battle of Britain Memorial Chapel, but is dedicated to the memory of all the aircrew who died while serving in the Biggin Hill Sector during the Second World War. There are eight major campaigns separately designated: Dunkirk, The Battle of Britain, The Dieppe Raids, The Pas de Calais Patrols, The Invasion of Normandy, Arnhem, The Ruhr, and the Crossing of the Rhine. But the Chapel is also in memory of the many who fell in the lesser campaigns, and in the more routine day-to-day operations of the war, the constant skirmishes that were not in themselves of enough significance to merit special recognition, but which collectively made a major contribution to the Allied victory. These men are also remembered.

The original Station Church was housed in a temporary building on South Camp – on what is now the Civil Airport site. As the course of the war progressed and the casualty lists grew longer, it became the wish of the Chaplain and station personnel to create a lasting memorial to their fallen colleagues. The claiming of the 1,000th enemy aircraft by Mouchotte and Charles in May 1943 acted as the catalyst in a plan to convert the Church into a memorial Chapel. The plan was the inspiration of the Chaplain, the Reverend Cecil King, who set in motion an extensive project which was very much a community effort.

A new altar was commissioned, flanked by oak-panelled reredos on which were recorded the names and the squadrons of the men from twelve nations who had died. And this list was complimented by an illuminated Book of Remembrance. The pulpit was made by a woodworker in the village, the astral crown and wings above the altar were carved by a Sevenoaks police constable, and the CO of the RAF Regiment airfield defence unit illuminated and lettered the pages of the Book of Remembrance. In this way the simple hutted church, typical of so many of its time, became something very special in the hearts of station personnel as a memorial to their fallen colleagues.

A special service was held on 19 September 1943 at which the reredos and Book

The Rev M H Edwards, RAF Chaplain-in-Chief 1940-1944, addresses the congregation at the Dedication of the Reredos and Book of Remembrance in the original Station Church on 19 September 1943. Group Captain Malan is seated with his wife at the right of the front row.

The altar and the reredos, showing the eight campaigns and the names of the 453 aircrew who were killed, listed beside their squadron badges. The reredos are flanked at each side by the national flags of the twelve nations involved.
(Photo: RAF Biggin Hill)

were unveiled by the Station Commander, Group Captain 'Sailor' Malan, and dedicated by the RAF Chaplain-in-Chief. The little Chapel was then in daily use as the Station Church until disaster came in December 1946, when it caught fire and was burned out. The cause was never really discovered and little was salvaged from the ruins, except the charred visitors' book in which one of the few names still legible was that of Churchill.

Padre King immediately devoted his energies to the building of a new Chapel – a purpose-built permanent building to serve as the memorial to all those young lives, and to signify thanksgiving for the deliverance of the nation and western civilisation.

In January 1947 the Air Council Committee on War Memorials announced plans to redesign and convert the old operations room, blitzed and evacuated in 1940, in to a new St George's Chapel of Remembrance. Mr W Wylton Todd, the architect who designed the original Chapel and who had been stationed at Biggin Hill, offered his services free of charge. The gallery of the operations room was to be included in the new Chapel, which was to be financed by Air Ministry subsidy and subscription from the relatives and friends of those who had been killed.

This plan was changed, however, and it was decided to build a completely new Chapel next to the main road on North Camp, on the site of an old aircraft hangar, opposite the Officers' Mess. The new building was to be financed through Air Ministry subsidy and public subscription. Padre King was still the driving force behind the new project and he launched a public appeal for funds which had the personal backing of Sir Winston Churchill, relatives and friends of the deceased aircrew also contributed and work was soon put in hand.

The appointed architects, Messrs A Beasley FRIBA, W S Harper ARIBA, and G A Williams of the Air Ministry Works Directorate retained a simple design. The exterior is neither ancient nor modern, finished in plain red brick with the large barn-style roof in heavy terra cotta tiles, the tower with a small roof to match. The red is rich in colour, and overall the building has a warm and pleasing, almost Mediterranean, feeling. Inside, the architects have retained the atmosphere of the wartime Church. Its severity and simplicity which meant so much to those who worshipped there, live on in the new building and are personified in the plain brick walls.

The Foundation Stone was laid by Air Chief Marshal The Lord Dowding, the wartime Commander-in-Chief of Fighter Command, on 25 July 1951, and by the autumn the Chapel was complete. The opening and Service of Dedication was conducted by the Lord Bishop of Rochester on 10 November 1951, and since that day the Chapel has been in full use as a Royal Air Force Church.

The Windows

The Chapel contains twelve notable stained glass windows from the studio of Hugh

The Station Window – one of a series of twelve windows from the studio of Hugh Easton.
(Photo: RAF Biggin Hill)

Easton, the designer of the RAF Memorial Window in Westminster Abbey. Each is to a common theme – *the cloud of witnesses* – and depicts the winged spirit of a young pilot embracing in his arms a badge. To the badges of the seven squadrons at the station during the Battle of Britain, have been added the badges of Fighter Command, 11 Group, the Station, and portraits of the immortal Spitfire and Hurricane, in badge form, depicted flying across the rays of the sun in a clear blue sky.

Easton was an Air Force man and flew Mosquitos in the war. With subtle discretion he has commemorated this by the inclusion of a tiny painted mosquito insect on the right-hand side towards the bottom of the Fighter Command Window.

Each window was donated – by Fighter Command, by RAF Biggin Hill, by Hawker Aircraft, Vickers Armstrong, Rolls Royce, Corporation of Lloyds and by private persons, and on most are small inscriptions and dedications in memory of colleagues and lost relatives. The 92 Squadron Window was provided by an anonymous donor and bears the inscription *And some there be who have no memorial.*

The Lectern

The lectern carries a fine gilded wooden eagle which was presented by the Misses Foster of Seal, near Sevenoaks, and the lectern Bible was the gift of 401 Squadron, Royal Canadian Air Force.

The Altar and Reredos

Four hundreed and fifty-three aircrew were killed whilst flying from the Biggin Hill Sector between 1939 and 1945, and they came from fifty-two squadrons. Their names are inscribed on the oaken reredos to each side of the altar, listed by squadrons and with the squadron badges beside them. The list is headed by the two Station Commanders who were killed in action. On each side of the reredos are the flags of the twelve Allied nations who mounted squadrons from the Sector.

The altar frontal is richly embroidered in traditional design and carries the RAF badge as its centre-piece. To each side and around this are the linked national emblems of the Allied nations and the frontal is headed by the quotation from Psalm 63.

The altar cross and candlesticks were donated to the Royal Australian Air Force by Sir Kenneth Luke of Melbourne for presentation to the Chapel in September 1965 – the 25th Anniversary of the Battle. They were designed to match the font.

The Font

The font was designed by the Reverend Vivian Symons, the one-time Vicar of Biggin Hill and the officiating Chaplain at the Chapel from its foundation until 1966. It is of aluminium and was made by Henry Wilkinson, a parishoner. Three

The entrance approach to St George's Chapel of Remembrance. The guardian Hurricane aircraft stands on the right and the Foundation Stone is set in the wall beneath the West Window. *(Photo: RAF Biggin Hill)*

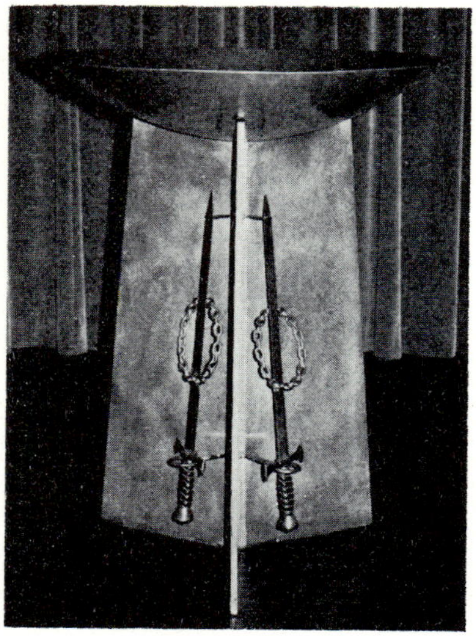

The aluminium font
by Henry Wilkinson ARE, ARCA.
(Photo: RAF Biggin Hill)

broad fins support the large bowl which is a fine example of aluminium casting. Between each of the fins is a sword encircled by a gilt chain – the emblem from the station badge. The sword blades are from saracen swords from crusader times but the hilts are replacements in the British pattern. Marshal of the Royal Air Force, Lord Tedder, presented the font on behalf of the Royal Air Forces Association, and it was dedicated in 1957.

The Book of Remembrance

The Book of Remembrance is secured in a display cabinet beneath the pulpit. It is a beautifully illustrated work and contains the names of the 453 who made the supreme sacrifice. The Book is prefaced by Spender's lines:

THE NAMES
of those who in their life
Fought for life
Who wore at their hearts
the fire's centre
Born of the sun
They travelled a short while
Toward the sun
And left the vivid air
SIGNED WITH THEIR HONOUR

The names are listed in diary form from January through to December. Each day the appropriate page is turned by the Orderly Officer, but thankfully there is not a page for *every* day of the year.

Commemorative Plaques

At either side of the sanctuary are plaques in memory of Polish and Canadian pilots who laid down their lives for freedom, and at the rear of the Chapel is a plaque of Delft tiles presented by the Protestant Churches of the Netherlands in gratitude for the hospitality given to Dutch airmen. In the gardens to the south of the Chapel is a plinth also presented by the Dutch.

The Bell

The bell in the tower weighs 11cwt and bears the inscription *Ring out the old, ring in the true*. It forms a link with St Mark's Parish Church of Biggin Hill – The Moving Church – for it was brought from All Saints Church, North Peckham, which was demolished by voluntary labour and rebuilt as St Mark's at Biggin Hill.

The Hassocks

The hassocks are in tapestry and were handmade to a common design by RAF and

WRAF personnel, by station wives and by ladies in the local community. The design incorporates the cross of St George and the sword and chain of the Station badge set against the silhouette of a Spitfire. Although of one design, there is of course variety in the stitches, and on many the embroiderer's initials have been included. The hassocks were completed in 1968 and were dedicated by the Bishop of Rochester in September of that year.

Recent Presentations

The Chapel continues to receive gifts and presentations from time to time, and significant recent additions include a silver chalice and matching paten, and a finely-carved second lectern.

The chalice and paten are particularly fine pieces. They are in solid silver and were commissioned for the Chapel by the American Eagle Squadrons' Association in 1976, in memory of the Americans of the three Eagle squadrons who lost their lives in operations from Britain during 1941 and 1942. The chalice is inscribed:

To those who came with us and now remain for ever

and the set was presented at a Service on 5 September 1976, during the Association's first reunion in Britain.

The second lectern was presented by the Belgian Air Force at a Service on 24 November 1976, and commemorates the Belgian aircrew who were killed in action with the RAF. It is a finely-carved wooden eagle in a style commonly found in Brabant in the eighteenth century, and it was made by craftsmen at the Belgian Air Force Apprentice School.

The Guardian Aircraft

The Chapel would not be complete without its attendant Hurricane and Spitfire standing guardian at its gates. They were dedicated as permanent memorials at a Drum Head Service on 19 September 1954 and they stand amidst beds of tall 'Peace' roses. The Hurricane, serial LF738, is a Mk IIc aircraft and was delivered off contract to 22 Maintenance Unit at Silloth in September 1943. It was used in the operational training role until August 1944 and it became an instructional airframe for fitters and mechanics a year later.

The Spitfire, serial SL674, is a Mk XVIe with the clipped wing tips. It is a post-war aircraft delivered to High Ercall in June 1945 and then used at Silverstone on operational training. In August 1947 the aircraft joined 501 Squadron at Filton, and after a flying accident in September 1948 moved on to 612 Squadron in April 1949. Its flying career then ended when it moved to storage at Cosford in July 1951 before being allocated to Biggin Hill for permanent display. The Spitfire was flown to the airfield by a Battle of Britain pilot, Air Chief Marshal Sir Denis Smallwood, GBE, KCB, DSO, DFC, who at the time was Station Commander at Biggin Hill.

The Chapel Today

Today the Chapel still serves as the Station Church. It is ecumenical between the Church of England and the Church of Scotland and Free Churches, and it is served by jointly-appointed officiating chaplains; the Vicar of Biggin Hill and a Methodist minister from the local community. Normal services are held each Sunday morning and on Battle of Britain and Remembrance Sundays special commemorative services are held.

Also, the building is open each day to the public and perusal of the visitors' book reveals that a constant stream of people comes from far and wide – from every corner of the globe – to pay their respect and pause and reflect in the quiet solitude of this now peaceful setting.

Padre King's wartime vision of a permanent memorial to those young members of his congregation is thus fulfilled – and his Chapel stands strong as we remember them today.

Commanding Officers

Major Orme
OC Wireless Testing Park Joyce Green – Moving to Biggin Hill in December 1916
1916 – August 1917
Major H B T Childs
OC Wireless Testing Park
Later Wireless Experimental Establishment
August 1917 – May 1918
Colonel L F Blandy, DSO
OC Wireless Experimental Establishment
April 1918 – December 1918
Wing Commander G P Grenfell, DSO
OC Wireless Experimental Establishment
December 1918 – November 1921
Major P Babington, MC
OC 141 Squadron
January 1918 – June 1918
Major B E Baker, DSO
OC 141 Squadron
June 1918 – June 1920
Squadron Leader I T Lloyd
OC 56 Squadron
November 1922 – September 1923
Squadron Leader Sir C J Q Brand, KBE, DSO, MC, DFC
OC 56 Squadron
September 1923 – August 1925
Squadron Leader F J Vincent, DFC
OC 56 Squadron
August 1925 – September 1926
Squadron Leader Elliott-Smith, AFC
OC 56 Squadron
September 1926 – January 1928

The Station was closed for rebuilding between 1928 – 1932

Wing Commander G B Dacre, DSO
October 1932 – January 1933
Wing Commander E O Grenfell, MC, DFC, AFC
January 1933 – February 1937
Wing Commander H G W Lock
January 1937 – November 1938
Wing Commander R Grice, DFC
November 1938 – December 1940
Group Captain F O Soden, DFC
December 1940 – June 1941
Group Captain P R Barwell, DFC
June 1941 – July 1942
Group Captain J R Hallings-Pott
July 1942 – December 1942
Group Captain A G Malan, DSO, DFC
January 1943 – October 1943
Group Captain H L Maxwell, CBE, DSO
November 1943 – February 1945
Wing Commander L Raphael, DSO, DFC
February 1945 – April 1945
Group Captain M W S Robinson, CBE
April 1945 – June 1945
Wing Commander E D Crundall, DFC
June 1945 – September 1945
Wing Commander C M H Outram
October 1945 – May 1946
Wing Commander G J Spence
May 1946 – June 1947
Squadron Leader G J Gray, DFC
June 1947 – December 1947
Squadron Leader G D Sise, DSO, DFC
December 1947 – May 1948
Wing Commander D C Smythe, DSO, GM
June 1948 – September 1950
Wing Commander A H Donaldson, DSO, DFC, AFC
September 1950 – January 1952
Wing Commander W Pitt-Brown, DFC, AFC
January 1952 – February 1953
Wing Commander D G Smallwood, DSO, MBE, DFC
February 1953 – October 1955
Group Captain J Barraclough, DFC, AFC
October 1955 – February 1957
Wing Commander P D Thompson, DFC
February 1957 – January 1958
Squadron Leader R S Salmon
January 1958 – March 1958
Wing Commander H W G Andrews, DFC
March 1958 – January 1959
Squadron Leader R S Salmon
January 1959 – February 1959
Squadron Leader E F Lapham
February 1959 – April 1959
Group Captain G P Seymour-Price, DFC
May 1959 – June 1959
Group Captain G N Amison
July 1959 – February 1962
Group Captain R B Morison, DFC, AFC
February 1962 – April 1962
Squadron Leader G N Bray
April 1962 – November 1962
Wing Commander R C Everson, OBE, AFC
November 1962 – November 1964
Wing Commander L F Wolsey
November 1964 – July 1967
Wing Commander I L McCombie
October 1967 – March 1970
Wing Commander P E Prior, OBE, AFC
March 1970 – June 1970
Wing Commander M Scholes, MBE, DFM
June 1970 – June 1973
Wing Commander R Gebbels
June 1973 – February 1976
Wing Commander K J O Balsillie
February 1976 – September 1977
Wing Commander A L Wright
September 1977 – April 1979
Wing Commander J R Myers
April 1979

Bibliography

RAF Biggin Hill Operations Record Book
Squadron Diaries and Operations Record Books
Fighter Squadrons of the RAF John Rawlings
RAF Biggin Hill Graham Wallace
The Hardest Day Alfred Price
Luftwaffe Alfred Price
The First and the Last General Adolf Galland
The Battle of Britain (Translated from Forces Aeriennes Francaises, October 1951) General Adolf Galland
The Battle of Britain Edward Bishop
The Fighter Pilots Edward H Sims
Bromley in the Front Line Lewis Blake
Royal Air Force Squadron Badges Peter C Smith
The Eagle Squadrons Vern Haughland
Luftwaffe Bomber Units 1939–1941 Jerry Scutts
Twenty-One Squadrons – The History of the Royal Auxiliary Air Force 1925–1957 Leslie Hunt
Personal Papers and Letters Some of the People

Acknowledgements

RAF Biggin Hill is grateful to the following, who by their generosity and assistance made the publishing of this book possible:

ABC photocomposition
Airfix Industries Limited
Armitage Shanks Group Limited
Barclays Bank Limited
Berney Baughen
BL Limited
British Aerospace
British Bata Shoe Co. Limited
British Petroleum Company Limited
Carreras Rothmans Limited
Thomas Cook Group Limited
Richard Costain Limited
CW Printing Limited
Kentish Times
James McNaughton Paper Merchants
Rolls Royce Limited
Josiah Wedgwood & Sons Limited
F W Woolworth and Co. Limited

The many people who were stationed at Biggin Hill in 1940 and who responded so generously to the appeal for material by sending personal memoirs, photographs and other items.

Ministry of Defence Air Historical Branch (RAF)
Ministry of Defence Adastral House and Whitehall Libraries